SpringerBriefs in Public Health

SpringerBriefs in Public Health present concise summaries of cutting-edge research and practical applications from across the entire field of public health, with contributions from medicine, bioethics, health economics, public policy, biostatistics, and sociology.

The focus of the series is to highlight current topics in public health of interest to a global audience, including health care policy; social determinants of health; health issues in developing countries; new research methods; chronic and infectious disease epidemics; and innovative health interventions.

Featuring compact volumes of 50 to 125 pages, the series covers a range of content from professional to academic. Possible volumes in the series may consist of timely reports of state-of-the art analytical techniques, reports from the field, snapshots of hot and/or emerging topics, literature reviews, and in-depth case studies. Both solicited and unsolicited manuscripts are considered for publication in this series.

Briefs are published as part of Springer's eBook collection, with millions of users worldwide. In addition, Briefs are available for individual print and electronic purchase.

Briefs are characterized by fast, global electronic dissemination, standard publishing contracts, easy-to-use manuscript preparation and formatting guidelines, and expedited production schedules. We aim for publication 8–12 weeks after acceptance.

Estevão Rafael Fernandes
Ana Karoline Nobrega Cavalcanti

Organ Transplantation and Native Peoples

An Interdisciplinary Approach

Estevão Rafael Fernandes
Department of Social Sciences
Federal University of Rondônia
Porto Velho, Rondônia, Brazil

Ana Karoline Nobrega Cavalcanti
Renal Transplantation Department
Rondônia State Department of Health
Porto Velho, Rondônia, Brazil

ISSN 2192-3698 ISSN 2192-3701 (electronic)
SpringerBriefs in Public Health
ISBN 978-3-031-40665-2 ISBN 978-3-031-40666-9 (eBook)
https://doi.org/10.1007/978-3-031-40666-9

This Springer imprint is published by the registered company Springer Nature Switzerland AG
The registered company address is: Gewerbestrasse 11, 6330 Cham, Switzerland

Paper in this product is recyclable.

To Alice

Acknowledgments

We, the authors, would like to thank the Federal University of Rondônia, the State Health Department of the State of Rondônia, and the Academic Department of Transplantation of the State University of Ceará. We also thank Dr. Tainá Veras de Sandes Freitas for her generosity and advice on this journey.

Finally, we also wish to acknowledge the financial support from the Brazilian National Research Council (CNPq; Grant 317075/2021-7).

Contents

1 Introduction: Thinking About Transplants from an Interdisciplinary Viewpoint 1
1.1 Getting Started 1
References 11

2 Looking Over the Shoulders of Giants 13
2.1 About the Method: Where to Start 13
2.2 A Little About the Comparative Method 17
2.3 Transplants in Native People: What Have We to Learn? 23
References 34

3 Native Corporeality and Organ Transplantation 37
3.1 Working with Other Cultures: What Now? 37
3.2 Talking About Health in Intercultural Contexts 38
References 52

4 Conclusion: Merging Horizons 55
References 57

Index 59

About the Authors

Estevão Rafael Fernandes is an anthropologist specializing in indigenous peoples of South America. He is currently an Associate Professor at the Federal Universities of Rondônia in Brazil. He is co-authoring *Gay Indians in Brazil* and *Queer Natives in Latin America*, both from Springer Publishing.

Ana Karoline Nobrega Cavalcanti is a Nephrologist, a full member of the Brazilian Society of Nephrology, having done her residency at the most advanced Brazilian hospital, the Hospital Israelita Albert Einstein, in São Paulo. Since 2018, she has been the nephrologist physician in charge of the renal transplant sector in the state of Rondônia, in the Brazilian Amazon.

About the Authors

Chapter 1
Introduction: Thinking About Transplants from an Interdisciplinary Viewpoint

1.1 Getting Started

1.1.1 Thinking Interdisciplinary

This is a sui generis text: it is born and thought from several frontiers and zones of interstice. The authors are a social anthropologist specializing in indigenous cosmology in the Brazilian Amazon and a nephrologist physician specializing in transplantation. Native cosmologies and biomedical knowledge are only sometimes recognized as possible dialogue zones in our academic corridors. We do not bump into each other in libraries, where our books are far apart, our epistemologies are entirely different, and we share different personal and professional ethos, most of the time.

However, not only this book is on a professional frontier between seemingly disparate fields. Historically the Rainforest is a space of meetings and mismatches between cultures and historical trajectories. We write from the Brazilian Amazon. In the state where we live, Rondônia, there are more than 70 ethnic groups and several traditional collectivities. We are on the border with Bolivia, a few hours from Peru, and live with a large community of Venezuelans. We feel and breathe Latin America and cultural diversity in our daily professional life.

In the encounters of this professional interstitial zone with this Amazonian mosaic, we find yet another possible frontier: the boundaries between health and disease. Moreover, this is where things start to get interesting. As will be seen, much of the effort in this first moment of our book will be to deconstruct the narrative that health and illness are distinct, well-established points operating according to the modern, Western scientific narrative. Giving a massive spoiler to this Introduction, we will insist on how this interdisciplinary look is fundamental to breaking ethnocentric notions of health, disease, corporality, care, life, and death. The truth is that

E. R. Fernandes, A. K. Nobrega Cavalcanti, *Organ Transplantation and Native Peoples*, SpringerBriefs in Public Health,
https://doi.org/10.1007/978-3-031-40666-9_1

Western scientific ideas reach a tiny portion of the world's population: white people, middle or upper class, living in large urban centers in the northern part of North America or central Europe.

Western medicine is a legitimate therapeutic path within a geo-localized perspective of science and knowledge (and, therefore, power). It depends on social class, formal education, and access to a well-structured social welfare network, among several other factors.

This challenge of humanizing medicine, in which there is an effort to listen to others in the science and health education process, is separate from the daily routine of most health professionals. In addition, this lack of interest, time, training, or personal vocation often generates failure in treatments that could save lives worldwide. If social factors affect and determine people's health, how to deal with culture as one of these factors? Is culture just a detail, or does it matter in the adherence and success of treatment for these people? What about the formulation of public policies? What about the structuring of services?

When it comes more specifically to transplantation, what cultural elements can be considered in the vast journey from initial approaches to successful treatment? Contrary to what the average citizen thinks, transplantation is not "taking the organ from one person and putting it into another." The procedure implies an approach that considers the impacts on those who donate and those who receive the organ, on their families, but also terms of consent, various tests, manipulation of fluids, and vital body parts. People with different trajectories and beliefs about their bodies and the bodies of others and these trajectories and beliefs become determinants throughout the process. How to count on someone's consent if the same task of explaining the procedures is challenging? How to deal with families if we do not know how family relationships operate? How do we explain test results to people who often do not understand what those various indices mean? How to help donors and recipients of organs received or donated by deceased persons deal with moral, religious, and intimate issues in these contexts? Transplants are not about organs but about people. That is the point here.

Nevertheless, what is the relevance of thinking about this topic in native contexts?

To put it briefly, Otherness—the concept we will see in this Introduction—pushes us to think about issues we would otherwise not consider. The encounter with the different affects and pushes us out of our comfort zone. In addition, the encounter with a radical difference takes us to places we did not even know existed. Native peoples offer a unique opportunity to force modern and Western thinking "out of the box," not as a humanistic matter but as a method. Comparing different realities allows us to contrast hitherto invisible elements. So even if you do not work with native peoples, reflecting on them (or rather, from them and their cultures) will hopefully raise questions about how the diversity of points of view can affect your professional practice—and this applies even if you do not work with transplants.

This book is, therefore, a way of trying to take you, the reader, out of your comfort zone, sometimes offering more questions than answers. It is also an interstitial book: written by professionals from different fields, in a Latin American peripheral context, about an epistemologically and humanly challenging theme. All you cannot expect in the following pages is a text that will confirm your points of view. On the contrary, our goal is to make you think from other interstitial perspectives.

1.1.2 "I Had Never Really Thought About It Before"

What is a body? The question may sound strange at first. The truth is that we do not think about it much.

We deal with bodies all the time, be it our own or others'. We expose our bodies on social networks at the beach, gym, or family moments. We teach our bodies how to dance or move in a certain way, at what times it should feel hungry, excrete, and express our feelings. There are tons of texts about bodies and how we deal with them, and this item will begin to problematize, precisely, this question—without pushing kilos of readers, at least at this point in our book.

However, the truth is that we have all been there at some point. We see a Brazilian on the street, and we are almost sure he knows how to samba. The authors of this book are Brazilians, and believe me when I say that this stereotype makes no sense to us. We see athletes on television and think, "My God, I could never do that," even though our bodies are fundamentally the same. Usain Bolt, Simone Biles, and Michael Phelps are made of the same material as we are, and yet we cannot do anything even close to what these people can do.

Bodies are not given but socially constructed—we will repeat this quite a bit here. This is not to say that bodies should not be understood as organic substrates but should not be read only in this way.

In Brazil, for example, it is expected in some indigenous ethnic groups for the man to go into seclusion after the birth of a child while his wife goes to work in the fields after giving birth [1]. In societies like the West, for example, this would be inconceivable, not because of some critical feminist view (at least not only) but also because of practical aspects. We treat childbirth practically as a disease, surrounding the mother with care and affection as if all societies in the world act or should act this way.

In addition, here goes the first essential concept. What Social Anthropology calls "Ethnocentrism." It is the perspective from which our world and our Culture function as the measure by which we judge all other Cultures [2]. The truth is that we do this all the time. People's view of those who are different is usually based on the assumption that their Culture is superior to that of others. History is replete with examples of this. The Greeks and Romans called themselves civilized, while others were considered barbarians—likewise, when Europe invaded the Americas. One of the justifications for the extermination of native peoples in the continent was a supposed, self-proclaimed moral and religious superiority of the invaders. Centuries later, empires would exterminate entire populations in Africa, Asia, and Oceania, adding economic and religious superiority to these justifications. Wars, conflicts, and genocides usually start from this vision of oneself as superior to the other and as a justification for territorial expropriation or murder.

However, there is a more subtle perspective to the ethnocentric view. It does not necessarily imply that one goes around killing others because he considers himself superior. Ethnocentrism distorts the Culture of the Other as part of the process of culture shock. It is a way of dealing with differences without thinking about our choices. Being Ethnocentric is a way for us not to think about how our Culture is just one, among several others. Neither better nor worse: just different. This is the perspective we seek to develop in this book, at least with the approach we chose to work with.

In this sense, when thinking about health, disease, body, and therapeutic itineraries in culturally distinct contexts, we stop having our options as the only possible ones, as if they were unique. This other way of dealing with culture shock also has a name: Otherness.

Alterity is not exactly a concept but a relationship. It is built on the recognition of differences.

It is about understanding that when we deal with ways of being, thinking, acting, feeling, and looking different from ours, we begin to rethink our alternatives. You do not think of yourself as a white person until you meet black people, nor do you think about your race privileges until you realize that other people are stopped by the police or followed by security guards more than others because of the color of their skin.

In Brazil, there are specialized police stations to attend to women victims of violence. The rates of femicide, harassment, and violence against Brazilian women are frightening [3]. However, some people think that a specialized police station attending specifically to women would be a privilege.

An interesting example of this happened a few years ago when we were having dinner at a friend's house. At the table, a friend of the authors said he was against this initiative. "I am in favor of precincts for people, without differences," he said at one point. The answer came as simply as it did quickly.

In what way could he, as a middle-aged, white, urban middle-class man, make value judgments in this regard? Since he would not know what it is like to be a woman victim of violence and to be attended only by men, with questions like "What were you wearing when you were raped?" for example. The discussion continued until the question was asked, "And if your daughter went through this, how would you react?" An awkward silence, followed by a "Yeah, put it this way ... I have never really stopped to think about it."

Moreover, it is this moment—the "I never stopped to think about it before" moment—that Alterity is made of. We believe that the options and paths we take are the only viable and best possible ones until we come across contexts and concerns that we have never been confronted with.

If, instead of killing native peoples, cultural bridges had been built, history would have taken entirely different directions. From this will come another example of what is Alterity, more related to the theme of this book. Alterity starts from the assumption that different cultures and social realities can provide original answers to problems plaguing us. One such issue concerns the notion of "individualism" as the counterpart of Modernity and Liberalism [4].

The Individual is seen, in the West, as an autonomous entity, morally and intellectually. It is the Cartesian subject, which thinks in the singular and is also in the singular. It is the "I," the measure of all things, a supreme and independent subject. It is not about "being selfish" or anything like that (in Latin languages, the concepts of individualist and selfish get confused). It is about having the "I" as the center of decision-making as if the field of possibilities in decision-making were independent of collective spheres or central power. It is the opposite idea of holism since people are interdependent, and society must be taken as a whole. Then there is 2020.

At the beginning of the Covid-19 pandemic, there was an illusion that people would support each other. The news showed images of neighbors singing in their windows, friends bringing food to people infected by the virus, videos and lives on social networks, and video-sharing sites with free concerts by famous artists. Everything led to the belief that the world would unite in a great wave of love for one another. However, shortly after, there were reports of threatening letters and notes to health professionals in their homes, looting of supermarkets, and a parallel market for vaccines. While the First World was getting second and third doses of vaccines, the third world was not even seeing the first doses arrive at its borders. The logic of the individual over the collective would be the keynote of much of the pandemic.

In Brazil, for example, an authority even stated, "Those who have to die will die," in a clear example of *necropolitics* [5] and a complete lack of empathy for the pain of others.

On the other hand, this behavior did not reach poorer sectors or traditional communities. The logic governing these groups' bonds is not individualism but holism. The notions of "person" and "individual" are not confused in these societies (we will discuss this difference later). The concern in small communities was from kinship and solidarity networks, not the dispute for resources. In these groups, there is an awareness that older people—and those susceptible to contamination by the virus—are the ones who keep traditional knowledge alive. In addition, in small groups, the kinship networks are more noticeable so that almost everyone is a cousin or cousins to each other. This meant their few resources during the pandemic were shared, and their solidarity increased. An example of this is that in several indigenous villages in Brazil, it was the natives who set up health barriers, preventing entry into their villages, given the absence of action on the part of the public authorities.

Communities that do not start from the paradigm of individualism have more consistent support networks, helping both in mental health and in welcoming people. Once you live with these holistically-minded people enough, the mistakes made during the Covid-19 crisis management become clear. The strengthening of economic alternatives that safeguard the livelihood of poorer families, the expansion of a universal and accessible quality health care system, the equitable distribution of vaccines, and policies to distribute masks to needy communities would only be part of the solution.

Alterity urges us to look away and rethink our own choices. A look at how these social groups dealt with this crisis could have provided powerful answers to the problem in other social contexts. As we will see, concerning the universe of transplants involving native people, this relationship between person and community becomes critical for health education, adherence to treatment, consent, and the day-to-day relationships with patients and the team when having native professionals involved. What is done to a member of these communities dramatically impacts the entire social group since the person is greater than the notion of the individual: it is permeated by various social, moral, and affective relationships culturally constructed.

Therefore, it is essential to step back and re-examine this.

1.1.3 When a Body Is Not Just a Body

Modern, Western, and Liberal societies learn to understand their corporality as a biological artifact. The narrative to which we are exposed from the moment we enter the school system is already well known. We are primates who, over the last few hundred thousand years, have had our cerebral cortex develop, come down from trees, and lost the layer of hair that used to surround us.

We gradually became *Homo sapiens* through a series of interactions between our Culture and environment. In a word: we evolved. Recently, especially after the nineteenth century, a good part of the decisions we make about our bodies as biological beings are based on the knowledge we have accumulated about how our bodies operate. We are educated, as scientists, to understand these processes objectively. Rigor and predictability are an inherent part of what we do. We are an organism, above all.

However, as already indicated, our goal is to broaden this—essentially organic—perspective of the body. It is the key word here, not to replace this view, nor to refute it, but to enlarge it.

It is fundamental to make an effort to understand that other views of corporality exist, which are as unique, complex, and legitimate as this biologizing view. We see the body as a biological entity because we have learned that this is the correct way within Western cosmology, but there are other ways of perceiving it.

In this sense, our body is not just an organism: it is an intersection of social relations and, thus, needs to be understood, also, as a cultural, social, and historical artifact. The body is not only an objective datum but also a representation, possessing a subjective character [6].

This is not surprising, either. We have a personal and affective connection with some bodies, seeing them as more than mere biological artifacts. Some bodies are, for us, different from others. This will vary according to the social relations from which we represent these bodies. Bodies are different stories, trajectories, and affections according to who feels or sees them. To say that bodies are biological structures made up of chemical and organic processes and the result of a selection and evolution process is to tell only part of the story. It is like saying, for example, that *Don Quixote* is about an eccentric fellow on a horse; or that *Hamlet* is about a son avenging his father. These would be true accounts in part, but completely superficial, incomplete, and distorted. What other layers are there if one wants to understand the relationship that human collectives have with their bodies?

In the previous item, we have seen concepts such as *Ethnocentrism* and *Otherness*, which is what the challenge is about here. Going beyond what we understand as body, individual, and scientific knowledge and opening our minds to other ways of being and being in the world, what do we learn? In addition, how can this learning be incorporated into our professional practices and epistemologies? In the following chapters, we will understand that some health education policies fail if they do not consider the audience they want to talk with. Science must be understood as a polyvocal space in an intercultural and subjective context. It is not about giving up scientific rigor; on the contrary, it is about opening up to the Other, making this science accessible and manageable in terms that are not our own. It is not about giving up accumulated knowledge but broadening this vision of knowledge by getting other knowledge, thoughts from different logics, and places of enunciation.

To put it briefly: even if you, as a professional, have a strictly pragmatic view of your patient's bodies, this does not mean they have that same view. Moreover, they are not even obliged to. In the world of transplants, this is evident on a day-to-day basis.

An organ is not, in these cases, only a set of cells and tissues. Patients and their families have emotional and subjective ties to what was, until recently, part of a loved one, for example. These bonds and relationships vary from culture to culture since the notion of corporeality, spirituality, individuality, and even the perspectives on life and death are also social representations.

As Vigarello [7] indicates, bodies should not be understood as something "natural," even though we tend to see them as "given." Bodies are, as he indicates, "fabricated" from the birth of the child until it becomes a Person. It is constructed as it operates in the context of gestures of sociability through bundles of social relations. Yes, we are born "Human" as an organic entity. Still, we become Persons to the extent that we are social subjects belonging to a given Culture and inserted in networks of social relations and kinship.

Opening here a brief and necessary parenthesis: *Human* refers to what we are as matter: bipedal, erect, etc. *Person* refers to what we are as a social construction: we are members of a specific Culture, socialized in a certain way and inserted in particular networks (ritual, kinship, religious, cultural, etc.) that make us members of that Culture and allow us to transit, socially, in it. *Homo sapiens,* on the other hand, refers to what we are as biological entities. In the latter case, this presupposes knowledge and belief in the biological explanation of our existence, setting aside explanations of a socio-cosmological nature and focusing on physical-chemical and environmental processes.

These themes will be duly worked on in Chap. 3, but this summary (very schematic and full of omissions) makes our central argument clear. The Body, seen from a social point of view, is not defined by the boundaries of individuality (a Modern and Western notion, as we have seen) but by the bundle of social relations in which it is inserted. It is not about seeing that body as Human, but as a Person, at the intersection of relations with its social peers. That body defines itself as Father, Son, Uncle, Grandfather, Godfather, Elder, Farmer, and Healer, all simultaneously. It does not exist as an "individual" because its perception of itself necessarily passes through the connections established between it and its cultural universe throughout its life. Thus, the body is an entity in permanent construction and flux since it is part of a bundle of historically and culturally established dynamic social relations. And it is from this "connection," this bundle of social, family, affective, and socio-cosmological ties, that the body is made.

It is not a natural object but an artifact, culturally constructed. This does not express, as De Castro [8] points out, "an alternative biological theory, and, of course, mistaken," but, on the contrary, manifests "a non-biological idea of the body." It is not a matter, and the author continues, of stating that our biologies are different, or "another view of the same body, but another concept of body, whose underlying dissonance to its 'homonymy' with ours is precisely the problem."

In fact, vast is the literature that accounts for how indigenous bodies are constructed culturally: the biologizing view treats bodies as sets of cells. In contrast, native cultures see them as bundles of relationships. How will this be explored in this book? That is what we will see next.

1.1.4 Some Clarification Before We Move On

Our main starting point is the understanding that the body is an axis of relationships and that it matters.

This point of view, from which the body is a cultural "apparatus," more than a mere biological "object," is synthesized by Rodrigues [9] as follows: the human body is not an individual but a cultural construction. Thus, it presents characteristics of cultural phenomena. It varies between people but also between cultures and contexts, historical and biographical. "Societies," he concludes, "construct bodies."

The above passage is in line with the central point of this research: native peoples have different (non-biological) views about their bodies, seen here as the result of the intersection of innumerable social and cosmological relations. It is not a matter, we repeat, of a "different view of biology" but of a different view of corporeity based on other notions of person, spirituality, kinship, etc.—something that certainly sounds counterintuitive regarding what is taught in Western health courses.

The Social Sciences show us, as already mentioned, that the binarism between Culture and Nature is a cultural heritage. It is based on modernity and Cartesian thought and sees Nature, the domain to which the human body would belong, as fixed and ahistorical. On the other hand, "Culture" would be a variable pole but prone to representations and subjectivities. This opens space for an ethnocentric view of the body, from which the prevalence of a pragmatic look on phenomena seen as "natural" would be the only explanation with truth value possible.

The fact is that most of the phenomena to which we are exposed in our daily lives do not relate to the realm of Nature or Culture alone: they are in an interstitial zone—we saw at the beginning of this chapter the importance of this fluid perspective. Even domains such as a forest are full of cultural representations, no matter how much one looks at them from a strictly academic point of view. Almost everything today is a hybrid between these domains: domestic animals, landscapes, and the weather. We will see below how, in native peoples, this dualistic vision between Nature and Culture is unjustified.

An example is Amerindian perspectivism [10], part of the "epistemological turn" in anthropology at the beginning of the twenty-first century. Formulated by the Brazilian anthropologist Eduardo Viveiros de Castro, it is about understanding a fundamental difference between Western thinking and native peoples. While some see Nature as unique and Cultures as multiple, others see Nature as numerous and Culture as unique. This implies a peculiar relationship with some groups of animals and with the dead, for example—a relationship that goes through corporality and its relationship with the native socio-cosmological universe. We will see about this in Chap. 3, but it is worth advancing the argument a bit, even if in a superficial and somewhat distorted way.

In short, since bodies are socio-historically situated artifacts, a sensitive perception of these other representations becomes fundamental to the intervention on them, which makes guidelines such as the ones proposed by this work so necessary. This sharp look to the difference that transcends the biomedical and objectifying look needs to be incorporated into health practices when it comes to procedures with native people or people belonging to traditional peoples.

These are highly complex cultural universes, and to reach the meanings these communities give to their bodies, deaths, diseases, and relatives is an effort to break our ethnocentric gaze.

To that end, this book will have the following internal division. In this Introduction, we seek to present thinking about transplants from an interdisciplinary viewpoint. As we have seen and will see in the following pages, human society is diverse in its view of body, health, illness, healing, and care. The Western biomedical gaze is not always prepared to deal with this other cultural universe when dealing with native peoples, often starting from an ethnocentric viewpoint. More than an organ or a body, in these societies, transplantation has its own social, cultural, and cosmological implications. Bodies are bundles of cultural relations in any human society, and any culturally sensitive intervention should start from this perspective. There is much to be learned from a look that takes otherness seriously.

Our next chapter will present what researchers and native organizations worldwide have produced regarding transplant-related initiatives. Doing an extensive literature review exercise, what has been done by transplant teams worldwide?

What are the positive experiences and challenges faced by these people? What do they report? In these national contexts, is there any specific legislation to deal with native peoples and organ transplantation? More than focusing on particular successful experiences, we will try to indicate what we can learn from what transplant teams worldwide are doing from a more generic point of view. We will also present texts that can help people interested in the subject.

Our second chapter will present approaches to Native corporeality from an anthropological perspective. What are the sociological implications of working with the body and health in other cultural contexts? What concepts and methods can be used to deal with these issues? This chapter will be about cultural and spiritual issues around organ donation and transplantation from the perspective of Natives. For this purpose, the authors will rely on a broad review of the anthropological literature. The idea is to present from fundamental concepts of the discipline—such as Culture, Relativism, and Alterity—to more mature discussions about Body, Person, Health, Illness, and Death in native contexts. We are conscious of the diversity of perspectives among native peoples, so we will seek to work from accounts of authors situated in ethnographically distinct contexts.

In our Conclusion, we will seek to tie these issues together, joining medical challenges in the context of transplants in native peoples with socio-anthropological approaches. As indicated here, the proposal is to present the theme from an interdisciplinary perspective, uniting theoretical and methodological horizons. Although there is a vast literature on Medical Anthropology and excellent articles produced on transplants involving native people, a more systematic socio-anthropological analysis of the theme is something to be done. It is essential to say that the book is aimed at those who do not have a background in social sciences and intends to address practical issues faced by healthcare teams, researchers, and managers.

However, we will still delve into important issues for understanding health and sickness processes in other cultural interpretation keys.

Since this is an introductory book, several issues that deserve further study will only be indicated, and it is up to the interested reader to seek knowledge for himself.

We will make notes indicating a bibliography whenever possible to guide you. However, we already know we will have some omissions precisely because we privilege an interdisciplinary and multifaceted look. Any generalization is a risk, but this would not be an excuse for not writing this text. Perhaps the community of health care workers and researchers will read this book and find it too related to the "Human Sciences." On the other hand, people who work with indigenous peoples may criticize several points here as overly simplistic. We are aware of these risks and hope to turn this book into a small series of others in which these issues can be deservedly and adequately explored.

Until then, we hope that readers can be indulgent with the authors and gain some benefit from this reading. Our goal is to provoke questions that may become, perhaps, more humane and culturally sensitive care for people of native peoples seeking health and transplant services worldwide.

References

1. L. Braga, V. Panem, Sobre seu viés de gênero entre os Zo'é. Mana **27**(2), e272201 (2021). https://doi.org/10.1590/1678-49442021v27n2a201
2. E. Baylor, Ethnocentrism, in *Obo in Anthropology*, (2012). https://doi.org/10.1093/obo/9780199766567-0045
3. R.L. Segato, P. Monque, Gender and coloniality: From low-intensity communal patriarchy to high-intensity colonial-modern patriarchy. Hypatia **36**(4), 781–799 (2021)
4. L. Dumont, *Essays on individualism: Modern ideology in anthropological perspective* (University of Chicago Press, Chicago, 1986)
5. J.A. Mbembé, L. Meintjes, Necropolitics. Publ. Cult. **15**(1), 11–40 (2003)
6. C. McCallum, The body that knows: From Cashinahua epistemology to a medical anthropology of lowland South America. Med. Anthropol. Q. **10**(3), 347–372 (1996)
7. G. Vigarello, Qu'Est-Ce Qu'Un Corps ? Catalogue De L'Exposition. Esprit (1940-) **11**(329), 210–212 (2006)
8. E.V. De Castro, O nativo relativo. Mana **8**(1), 113–148 (2002)
9. J.C. Rodrigues, *Os corpos na Antropologia. Em: Críticas e atuantes: Ciências sociais e humanas em saúde na América Latina* (Editora Fiocruz, Rio de Janeiro, 2005)
10. E.V. De Castro, Cosmological deixis and Amerindian perspectivism. J. R. Anthropol. Inst. **4**, 469–488 (1998)

References

Chapter 2
Looking Over the Shoulders of Giants

2.1 About the Method: Where to Start

2.1.1 Two or Three Words About How These Concerns Were Born

A few things need to be said about the genesis of this book before moving on.

The authors of this text, an anthropologist and a nephrologist specializing in transplantation, are probably the only people in Brazil who research transplants in indigenous people in Brazilian territory. For the first time, a systematic effort is made to understand the consequences of transplantation involving indigenous people. Some of the thoughts in this book reflect the anguish of not having a starting point for consideration—and, consequently, for action. Where to begin is the concern of many colleagues who, like us, are in regions that are not accommodated by advanced medical systems or whose routines do not allow for systematic reflection on their practices. It is a trajectory of absences. We will begin this chapter by sharing these anxieties and how we have tried to respond to them.

Given the absence of Brazilian literature on this theme, our first move was to search for texts reflecting on Native American peoples in Latin America. To our surprise, we found a need for published systematic studies on transplants in indigenous peoples in Latin America, in contrast to the vast literature available in Anglophone countries. We'll talk more about that on the following pages, which explains why we thought of writing this book in the first place. We knew what we wanted to understand but didn't have a starting point.

Speaking of Brazil, a country known worldwide for its ethnic diversity, especially in the Amazon, some findings are shocking. There are no systematic data on how many or which indigenous people have been involved in transplants, even with a particular indigenous health subsystem. The ethnic component is ignored by

E. R. Fernandes, A. K. Nobrega Cavalcanti, *Organ Transplantation and Native Peoples*, SpringerBriefs in Public Health,
https://doi.org/10.1007/978-3-031-40666-9_2

transplantation systems, with natives treated as "*pardos*" (a Brazilian expression that characterizes mixed-race people, literally "brown people").

This causes their cultural specificities to be lost throughout treatment, compromising conduct adherence and its success. There are no specific health education and communication actions for the natives who use the service or the health teams that provide this assistance. Likewise, there are no protocols regarding consent, manipulation of fluids, tissues, or images, or providing interpreters or dialogue with traditional knowledge. Everything happens as if more than one million Brazilian indigenous people, distributed in almost 300 ethnic groups, had no culture or therapeutic paths of their own.

Brazil's lack of systematic reflection has led us to focus on other countries where literary production is more consolidated [1, 2]. The comparative method is relatively common in the Social Sciences, and such topics are necessarily interdisciplinary [3]. The comparison allows us to merge horizons with other social realities and cultural and political configurations, isolating elements and starting from somewhere. The perspective that knowledge comes from accumulating inside and dialogue with peers is commonplace in the Sciences. In this sense, the extensive production in other countries can illuminate the lack of more systematic reflections on Brazilian reality—and this is where something interesting arises, on which this text will focus.

Throughout the research, mainly using various online databases, there were several references to cultural aspects related to transplants in native peoples, whether in health communication and education, information, or public policies. New Zealand and Australia, followed by Canada and the United States, concentrate the specialized literature, with hundreds of texts on the subject produced by professionals from various fields and based on multiple approaches and themes.

However, when following the same path in the databases, surprisingly, there is a lack of articles written by Latin American researchers on native peoples in Latin America. Articles in Spanish and Portuguese that raise issues related to transplants involving native people, their anthropological and health implications, analysis of specific policies or protocols, or health education actions: none of this exists systematically when considering native peoples in Latin America.

There are some hypotheses for this absence.

First, health services are only sometimes effectively structured to consider the cultural aspects of services offered to indigenous peoples. Even though there is often a bureaucratic-administrative structure in this regard, there is a lack of practical awareness-raising actions for teams that attend to native people in medium and high-complexity services. Even if the groups that provide first aid in the villages have this sensitivity, alterity is lost along the way, with the Western biomedical model prevailing over traditional care and reception paths [4].

Second, in some of these countries, indigenous peoples are in the midst of an epidemiological transition, with health services failing to anticipate the increase of diseases such as hypertension, obesity, and diabetes in native populations as a future demand in the field of transplantation. The increase in chronic diseases is often viewed as a result of an eventual "loss of culture," generating prejudices and blaming the victims, causing the service to not adapt to these demands and focus on acute or endemic diseases.

It is important to remember that in several regions of Latin America, contacts with native tribes have intensified, especially in the last 80 years, given the developmental agenda in the area after the Second World War. This means that only recently, there has been an adult population whose effects of prolonged contact, sedentarism, changes in diet, and sanitary conditions are felt more intensely. In practice, indigenous movements seek to address more immediate demands in their struggle for territories, education, sustainability (environmental and economic), and health [5]. However, needs in the health field often seek solutions to more immediate problems, such as epidemics, vaccines, nutrition, or water and soil pollution caused by invasions into their traditional territories.

Furthermore, unlike Anglophone countries, most native peoples in Latin America are not in large urban centers but in small towns in rural or isolated areas where more complex health services, such as transplants, are non-existent or practically inaccessible.

Finally, even though Latin American countries are cultural melting pots, their Constitutions and legislation need to understand multicultural perspectives. Thus, external institutional actors design and manage public policies, often unaware of indigenous demands or with little dialogue with these communities. Moreover, with the rise of the extreme right parties in some of these countries, part of the budget for assistance to indigenous peoples has been withdrawn [6], favoring the interests of logging companies, farmers, miners, or missionaries.

In other words, the lack of a more systematic analysis in the literature regarding transplants among native peoples in Latin America should be viewed in a broader context. It is not a matter of lack of interest but of a series of structural and historical reasons that lead to a neglect of these demands both within the indigenous movement, which struggles for primary conditions of sustenance and survival, and by the National States of the region, in which more immediate demands such as acute diseases are prioritized and health is seen, almost always, apart from issues such as territory and education.

In addition, public health services in these countries also face financial difficulties and suffer from poor management [7]. The problem of the need for a specific public policy for transplants for indigenous peoples encounters the same issues that affect the lack of these services for poor or rural populations. The lack of equity in health care is not only a problem for native peoples but for the poor population and minorities in general. The ethnic issue becomes just one more element in the field of the social determinants of health, in general.

The question then becomes: how to break this cycle?

One possible alternative is to highlight its existence. Since no robust data demonstrate the need for these policies, managers and politicians do not see it as a demand. In this sense, the role of researchers is fundamental as a way of unveiling the often exclusionary nuances of national health systems in these countries. The creation of multidisciplinary research networks with the international collaboration of these researchers is also an exciting alternative, including as a way of seeking funding for the research.

Another element is the formulation, together with health managers, of projects in health education with indigenous organizations, health councils, and institutions and actors responsible for health social control. Often, native people do not know that they have access to transplantation as a form of treatment, so how this information reaches the villages, considering native cultures, can be a differential in the medium term.

Finally, including disciplines with a multicultural perspective in schools and medical faculties in these countries is an urgent task. Medicine is a course geared toward economic elites in several Latin American countries. These students come from urban centers, and many unintentionally reproduce a lack of knowledge about topics related to cultural diversity. They are white, upper-middle-class individuals who have never had contact with alterity, either in their trajectories or in their formative paths. This cycle of invisibility of topics such as race, gender, and ethnicity in medical courses must be responsibly addressed as an essential part of this formative journey rather than just a curricular requirement.

The regional disparities affecting medical education and healthcare for underprivileged populations reflect how these states deal with their indigenous and poor or rural people. As this cultural sensitivity toward complex health topics is gradually addressed, there is a tendency for the production of the subject to gain ground in Latin American countries. Interdisciplinary barriers will be overcome, and in the process, much of the institutional racism in these countries will be addressed, as the structural methods upon which this racism is based will be revealed. This book is a first effort in this direction. It is also an invitation to form a Latin American collaboration network on transplant among native peoples on the continent, if possible consistently with the participation of indigenous professionals, and a challenge to think about how to offer these services from a dialogue that is genuinely welcoming from the standpoint of Latin American cultural diversity.

The previous paragraphs are a remarkable synthesis of what we will see in the following pages of this chapter. This grand synthesis explains the lack of a closer interlocution with the Latin American context, despite the extensive ethnographic knowledge of the continent's peoples. We are aware that this is also the case in other geographical contexts (especially Asia and Africa).

In this sense, it also explains why we opted for the comparative method—to be presented in the following items. Not only is there a discrepancy between transplants among white people and minority groups, but there is also a relative scarcity in the literature that questions this gap. Similarly, there is a wide gap between what is produced in English-speaking first-world countries with the other continents. As we explained above, this contrast should also be the object of more systematic research.

Finally, several aspects discussed in the previous paragraphs will be revisited below for a simple reason. Although the comparison is a well-established method in the social sciences, the same is different in health research.

Epidemiological studies tend to have, traditionally, a quantitative approach. Thus, systematic reviews based on strictly quantitative aspects could be expected from texts like this. And this is where interdisciplinarity comes in. The way the literature has been addressed and incorporated by us, the authors, is to keep the challenges of thinking about transplants in the context of Alterity at the forefront. When conducting a more formal survey of the texts, we felt something was lost in the readings. We were constantly reminded by the various articles we read about how important it is to flex the listening in topics like this. To filter out the efforts of those who thought about the topic merely from numbers and statistics would sound cynical and contradictory.

This chapter aims to bring a general and humane view to transplantation thought in native contexts. We promise an interdisciplinary book considered interstitial zones. Several of the points brought up by the authors read—and recovered here synthetically, as far as possible—will be a bridge to the next chapter, more ethnographically motivated.

2.2 A Little About the Comparative Method

2.2.1 Some Concerns

First, it is essential to understand the size of the problem we are addressing here. As we will see in the following items, native peoples are historically excluded from other social groups regarding transplants. Not only is there a great deal of pent-up demand, but there are fewer organ offers, a longer wait, and structural barriers in their care.

In the United States, for example, the Centers for Disease Control and Prevention (CDC) estimates that chronic kidney disease (CKD) and end-stage renal disease have a prevalence 1.4 times higher among Native Americans than among non-minority populations, as Anderson et al. [8] point out. However, as the same authors indicate, despite having mortality rates caused by kidney failure 2.5–3.5 times higher than those reported for non-Natives, Native people are less likely to receive transplanted kidneys.

Also, in this public, there seems to be underreporting and a lack of more narrow and culturally sensitive care. In this sense, complementing the data from a study conducted with urban Indians in the United States, it is observed that their waiting time for a kidney transplant is 40% higher than that of non-Indians, even though terminal kidney disease has an incidence 3.5 times higher among natives and with an average age of 6–7 years younger than whites [9].

This context denotes an increase in the demand for health care for these individuals in spaces not always directly or adequately served by the Health Systems, making it difficult to provide adequate care. It is not only about overcoming possible language barriers but also about accessing another cultural universe based on other processes of care, healing, sheltering, health, and illness [10–14].

Remember that the very concept of "health," proposed by the World Health Organization (WHO) as something beyond the absence of disease, but a "complete state of physical, mental and social well-being," seeks to account for this scope, understanding the processes of health and illness also within their subjective, cultural and socially constructed and recognized spheres. From these points, when it comes to transplants in native populations, specifically, there are several possible problematizations.

First of all, having a vision from a multicultural perspective is necessary. Each of these peoples represents its cultural universe and constitutes a unique reality. How these people conceive life, corporeality, and eschatology implies challenges in the approach to donors and recipients, for example. There are barriers in these cultures that must be considered when it comes to receiving organs from living or deceased donors. When studying the cultural influence and the desire to donate among Native Americans in the United States, for example, Bongiovanni indicates several socio-cosmological factors that influence the intention of these subjects to donate their organs: a spiritual connection with the organ recipient; the belief that without any organs in their body, their spirit will not pass to other planes or will not rest; among other reasons were pointed out by the interviewees [9] as critical spiritual elements in their cultures. Thus, this concept of health should start from a holistic view, comprising emotional, social, mental, and spiritual aspects, transcending a strictly biologizing view. We have seen something along these lines in our Introduction, and the following pages (and chapters) will go even deeper into these aspects.

Second, many of these communities do not have access to more information regarding transplants precisely because of the lack of a specific public policy. Working with Australian aborigines, Anderson draws attention to how important effective health education and quality communication are to break a cycle of uncertainty about the causes of kidney disease and the commitment to treatment in this population [15]. In fact, by proposing a comparative look based on international experiences in the field of transplantation involving native people, we seek to expand the sphere of possible potentialities in the area of action based on mature reflections and consolidated experiences in the context of public policies.

However, to indicate the need for specific health education and communication actions is not only to point out the gap of those actions aimed at indigenous people but also for the teams that serve this public. If, even among the non-native population, the non-adherence to donation and transplantation often comes up against misinformation, it is plausible to assume, based on the sources consulted, that an approach with native people should also consider their culture and social organization. Although access to lists and consultations is, in principle, universal, this does not mean that this information is accessible—socially, linguistically, and culturally speaking—to these people, nor can they handle it.

Such experiences show that it is possible to face challenging issues such as prejudice, fear, and racism—as well as make clear the set of barriers to be overcome from a legal, bioethical, and cultural point of view.

Thus, there is an apparent dilemma: is it possible to think of a transplant guideline that considers alterity but does not abandon the minimum medical precepts? Where to start? Indeed, several readers are concerned about maintaining a certain pragmatism from their biomedical education. Minimum criteria are observable in tests and examinations of blood, tissues, compatibility, and immunity. This is not a request to give these up. However, other cultures see these processes differently (this will be explored in the next chapter) based on their own conceptual and cosmological keys. Stages related to treatment adherence, consent, and family and community participation in the patient's improvement necessarily pass through the understanding of this other interpretative key—the holistic view to which we referred in the Introduction.

Subjectivity becomes essential in doctor–patient relationships and health care and attention policies. A sensitive look, culturally speaking, matters not only to imply the humanization of treatment but also to increase its chances of success. This sensitivity engages staff, family members, and patients in peripheral contexts, such as ethnically differentiated ones.

As we saw in the Introduction, the view that an organ is merely a collection of cells and tissues simply does not fit: patients and their families have emotional and subjective ties to what was, until recently, part of a loved one or themselves. In addition, these bonds and relationships vary from culture to culture since the notion of corporeality, spirituality, individuality, and even the perspectives on life and death are social representations. The body is a bundle of social relations, not a natural artifact, as we have tried to indicate since our first pages.

It is possible, then, to create reflections with practical effects, but at the basis of which is respect for the cultures and health-disease processes of these people and their communities.

What are the issues involved in thinking about transplants involving indigenous people? From a theoretical point of view, interdisciplinary perspectives intersect: it is impossible to have a minimally responsible approach without considering the social implications of a problem that transcends medicine, epidemiology, or public health, for example, as if they were distinct and watertight fields of knowledge.

A clear example of this occurred a few years ago.

In its November 11, 2001 edition, *O Estado de São Paulo*, one of the most important Brazilian newspapers, carried out a lengthy report on the "Xingu Project," an indigenous healthcare program in the Xingu River basin in central Brazil. Whoever had in his hands that Sunday edition would read a text written by journalist Luciana Miranda, reporting how changes in diet and lifestyle caused by long coexistence with nonindigenous society now caused problems such as obesity, hypertension, and diabetes. According to the specialists Miranda heard, even in treating these diseases, the shamans should be listened to and the traditional knowledge respected.

At the end of the report, one reads something that catches one's attention for reasons to be indicated further: "It was thanks to the integration between doctor and shaman that a health team performed the first kidney transplant in a native in December 1996. Dombá, 40 years old, had a malformation of the urinary tract, a congenital disability." Over time, his kidneys malfunctioned until he had to leave

Xingu to undergo hemodialysis in São Paulo. After that, he needed a transplant. "Putting the organ of another person inside the body of a Native is a delicate matter," says Carlovich [Jorge Carlovich Filho, coordinator of the Clinic]. The surgery was only performed after the shaman analyzed the case. Dombá lived for four more years with the new kidney in his village in Xingu before he died. "Dealing with cultural differences and not being the owner of the truth is the constant task of the health professional who works with the native."

Dombá Suyá (the last name refers to the name of his ethnic group) reported, after the surgery, having seen his shaman—whose physical body was in the Xingu—during the procedure, which would have reassured him.

There we have several interesting elements indicating precisely the delicacy of the theme highlighted in the excerpt above. Although they have already been cited here, it is worthwhile to resume them through synthesis as premises of this work.

First, the epidemiological impacts of contact with non-native society. In this sense, for some years now, several researchers have been drawing attention to the state of epidemiological transition observed in these communities, "in which, although infectious and parasitic diseases persist as important causes of death, there is also a significant weight of chronic noncommunicable diseases and injuries, poisoning, and external causes" [16]. With increasing access to salt, sugar, fatty foods, and, in many cases, a greater sedentary lifestyle, there has been a growing number of indigenous people with problems related to diabetes, obesity, and kidney disease, for example.

Moreover, the practice of "placing another person's organ inside the body of an indigenous person" is sensitive, as mentioned in the report, to the need to think from an intercultural perspective of health, disease, body, death, and person. Vast is the literature in the field of epidemiology that urges us to build a plural look at these processes: Trostle [12], Hahn [17], and Helman [14], for example, provoke us to think about the relationship between Culture and Medicine, adjusting our gaze to the differences, without exoticizing them. It is about understanding that different peoples have different explanations for their illnesses, their healing processes, and their concepts of health, death, and care.

Dombá's mention that he saw his shaman on the operating table cannot be reduced to delirium or an eventual placebo effect, but, on the contrary, is illustrative in this sense: his therapeutic itinerary necessarily implies having his shaman present as a guide, counting on his presence, even if on a spiritual plane.

Finally, these other notions of corporeality, health, death, etc., represent a challenge for the health teams, from those that provide the first care, still in the village, to the one that performs the transplant. There is a delicate balance required: if, on the one hand, there is the need for an intervention based on well-established techniques and protocols in terms of performing exams, handling blood and other fluids, for example, on the other hand, there is a series of rights to be considered in this care, especially respect for the patient's culture—in this case, an indigenous person.

2.2.2 *Comparing: Why Is This a Good Option?*

The effort to open up to these other representations of corporality directs the inquiries of this book to the broadening of the concept of "body." Native peoples have different (non-biological) views about their bodies, seen here as the result of the intersection of innumerable social and cosmological relations. It is not, we repeat, about "another view of biology" but another view of corporeality based on other notions of personhood, spirituality, kinship, etc.—something that certainly sounds counterintuitive regarding what is taught in health courses worldwide.

In this sense, as Rodrigues indicates, "against the Cartesian ethnocentric dualism of our intellectual heritage, let us never forget that 'biological' and 'cultural' are just concepts. The human body does not have two sides—one fixed and biological, the other variable and cultural—but only one" [18].

In short, given that bodies are socio-historically situated artifacts, a sensitive perception of these other representations becomes fundamental to their intervention of them. This keen look to a difference that transcends the biomedical and objectifying look needs to be incorporated into health practices when it comes to procedures with indigenous people and belonging to traditional peoples.

An example of this is what Britt et al. write when they propose community-based participatory research to engage the Native American community in health promotion, knowledge, and health education, promoting kidney donation and transplantation. Initiatives like this do not seek to impose the biomedical view but to exchange experiences and learning with the native population to "develop strategies for prevention and health promotion that serve the community" [19].

Like this one, several successful official initiatives are being developed or underway in other countries to think about transplants for indigenous peoples.

One example comes from the Office of Minority Health of the US Department of Health and Human Services, which has a website specifically about organ donation for Native Americans in the continental United States and Alaska. In Canada, in turn, there are cell phone applications aimed at First Nations people, such as the Kidney Check program or initiatives to discuss how the health system can welcome Indigenous patients and their families in actions related to organ transplants—some of them promoted in collaboration with the Network Environments for Indigenous Health Research National Coordinating Centre, the First Nations and Métis Organ Donation and Transplantation Network, and the Can-SOLVE CKD Network. There are also initiatives such as Hope and Healing, a campaign for organ donation through the use of social networks in the United States, or the IMPAKT Program (Improving Access to Kidney Transplants), whose purpose was to investigate the problems in the greater adherence to kidney transplants by Aboriginal Australians.

The research on which these data are based is the so-called integrative review in the terms proposed by Souza, Silva, and Carvalho (2010). According to these authors, the integrative review would be a broader methodological approach to reviews, allowing the inclusion of experimental and non-experimental studies for a complete understanding of the phenomenon analyzed. In addition, this type of review would combine data from theoretical and empirical literature, incorporating the definition of concepts, review of theories and evidence, and analysis of methodological problems of a particular topic (p. 103).

As they indicate, this type of review would have six phases: (1) elaboration of the guiding question; (2) literature search or sampling; (3) data collection; (4) critical analysis of the included results; (5) discussion of the results; and (6) presentation of the results. The integrative review is presented at the end of the "reduction, exposition and comparison, as well as conclusion and verification of the data" [20]. In other words: there is no easy way out or magic formula. Much of the filter depends on the researcher's intention, interest, and background. It is the interpretation of texts, not just data, a work of intellectual mining in its simplest and rawest form.

Many articles are available—including reviews, case reports, original articles, etc.—from the experiences of researchers and health professionals in English-speaking countries (mainly Australia, Canada, the United States, and New Zealand). The paucity of data on Latin America was striking, which explains the long explanatory note at the beginning of this chapter.

The fact is that any search made in databases (such as the Virtual Health Library and PubMed, for example) using terms like "indigenous peoples" + transplantation; (aboriginals) AND (transplantation); "native American" transplantation; (first nations) AND (transplantation); and (indigenous peoples) AND (transplantation) quickly reaches more than 400 published articles. Each of these expressions has regional relevance: Native American is the most used in the United States, while First Nation is most used in Canada and Aboriginals in Oceania. However, the terms are not mutually exclusive, and more than one term may be used in an article—for example, a text written in Australia may use "indigenous peoples" as well as Aboriginals, depending on the type of discussion. Also, not all texts are about transplants and indigenous peoples from intercultural concerns or in the field of public policy. Several are about dialysis or diabetes and kidney health, for example. Thus, there follows an exhaustive—but rewarding—reading of all the summaries of these texts.

The text's writing date was not incorporated as an inclusion or exclusion criterion, given the nature of the theme: these are reflections built and consolidated little by little and in parallel from the authors' experiences, without any rupture of paradigms, protocols, or approaches over time. When systematizing the data, the division into countries was chosen, as it is understood that such a view allows for a broader understanding of the national contexts, separately comprehending the structures of the health systems in each state. From there, the themes can be unfolded to public policies or health communication, for example. This is where the comparative method comes in.

In proposing here the comparative method, we start with a practical question. The vast majority of countries in the world do not have specific literature on transplantation in native peoples. It is sparse and unsystematized, except for some English-speaking countries or nearby nations. Thus, our contribution in presenting the comparative method is to provide an interpretative key to help researchers develop a method of literary analysis that can help them within their national context.

Thus, in this effort to organize these readings, the comparative method was used to compare contrasting cases. The comparative method compares different worldviews, privileging experience, and cultural contexts, elucidating different horizons.

On the contrastive approach in the comparative method, Olivier Giraud writes, "The comparative strategy of the most contrastive cases has the simultaneous goal of isolating cases that are representative of a diversity of situations and working with cases that present clear configurations that are conducive to convincingly applying tests" [3].

By broadening our gaze, new perspectives open up. That is, they seek to reveal factors that, by themselves, would not be easily visible or isolated. As this author indicates, comparing contrasting cases would have several advantages, such as questioning naturalized mechanisms in-depth and allowing a deep analysis of the cases.

Thus, looking at these other national contexts is justified to compare them with local situations. What can be learned from these experiences to propose approaches appropriate to the context of a given country? Is it a different legal framework? A concern with multiculturalism? A process of training health professionals? Of the structure of health care policies for native populations? What approaches and strategies can assert these populations' right to differentiated health care while ensuring effective treatment, engagement, understanding, and active participation in the care process?

The comparative method from a contrastive approach allows this type of questioning, not guided by the lack and precariousness of services in a given place, but by a propositional look. How can we think of this model based on what already exists in other realities in the world, the care and reception of indigenous demands in transplantation?

In other words, we will seek to organize the material found in the review to have a comparative look that informs the subjective processes of illness, healing, death, and care and those in the public health policies aimed at these indigenous populations. In what ways can these two universes dialog, and, in fact, do they dialog based on experiences going on around the world? What can these experiences inform and teach us about what and how to do?

With these questions in mind, we analyzed the articles found, arriving at some important points to be presented. Since there are hundreds of essays, it is impossible to cover all the points raised, so we will synthesize what most caught our attention regarding models that can be extrapolated to other national contexts—or improved, even in countries with vast systematized literature.

2.3 Transplants in Native People: What Have We to Learn?

2.3.1 What Is to Come

The stages of transplantation vary somewhat depending on the type of intervention and region but generally follow the same basic structure.

Some deficiency is detected in the patient, and his demand is referred to a specialized team. From then on, they start to be seen as an eventual organ receptor. The transplant team is determined according to the patient's place of residence and choice. It is up to this team to analyze treatment needs and possibilities. It is up to this team to define, also, the viability of a transplant.

In the meantime, the recipient's data becomes part of a computerized registry, and if the transplant team agrees, they will be placed on a waiting list to receive an organ. While he is undergoing various and sundry tests, both he and his family are duly informed about what a transplant is, its implications, and its consequences, and it is up to the patient to consent to the procedure.

Once the computerized system crosses donor and recipient data, more tests are done to ensure compatibility between donor and recipient. All this is always done with a professional team that supports them at each stage. Once the organ is available, the system defines who is the most suitable recipient to receive the donation. The clinical situation of the recipient is systematically evaluated and re-evaluated before the actual surgery takes place.

After the procedure, the recipient continues to be accompanied by a team that ensures that the new organ will have a lower chance of rejection. In addition, this team is always responsible for guiding the recipient as well as his family in relation to eventual changes in habits and medication to be taken by the recipient throughout his life.

The point is: even if formally all these steps follow strictly objective criteria, there is room for subjectivity of the actors involved in several of these steps. The literature calls our attention to this, and deconstructing the notion that all this is linear and pragmatic is essential to move forward.

The first step, for example, is to detect the demand, assess the patient's medical condition, and refer the patient for treatment. As we will see, one of the problems faced starts there. Native populations don't always have a continuous health care system, with teams counting on medical specialists with enough time to detect this kind of problem and need. Most native peoples live in total or relative geographic isolation, which greatly hinders both the assessment of the patient's condition and the referral of the patient to a specialized transplant center.

In fact, the very process of educating that health community about what a transplant is, in itself, a real problem. We will see a limiting situation in the next chapter that will problematize the view that the Western medical view is intuitive. Health communication often comes up against the challenge of allowing itself to be reinterpreted by the population for which it is intended. In addition, there are already problems at this stage concerning the very handling of bodily substances. Blood itself, for example, is often related to important cosmological aspects for native groups, and its manipulation is something to be done carefully and in light of the possible political and bioethical repercussions for the population and the medical team [21].

The same goes for the pre-transplant stage. Both the tests and the consultations with specialists must be done, taking into account the cultural repercussions of the procedure. The fact that this is a native population with its own social and socio-cosmological dynamics should not be treated as a mere detail. This is the starting point because it is from their own point of view that the patient starts to understand the risks and benefits of the procedure they will undergo. Remember what we have already seen here: placing the organ of one person in another is not a simple task and requires cultural sensitivity on several levels.

Even the allocation of the recipient on a waiting list is somewhat problematic. The second part of the literature, as we will see below, calls attention to the fact that even algorithms that define the waiting list for organs tend to disadvantage ethnic and racial minorities.

Finally, if the organ arrives, there are other variables. The geographic isolation of the native community can hinder the receipt of the organ in a context where time is critical. If the organ comes from a close relative or a deceased person (or, in the case of native donors, if the organ is taken from a deceased person), there are innumerable religious and cosmological repercussions. The Maori example given below is evident in this sense. Whom the organ comes from or goes to, and under what circumstances, matters in the end.

In the case of Dombá Suyá, we see how the surgery itself also has religious consequences for native cultures. Integrating native knowledge in the preoperative period can be an interesting strategy to guarantee not only the tranquility of the recipient and his family but also the adherence to treatment. Lévi-Strauss, for example, already in the 1940s wrote texts now considered classics to situate the importance of the manipulation of symbols in healing and death processes in native contexts [22]. Again, the idea of several strangers manipulating your open body in a cold environment may cause strangeness and even repulse in several cultures, used to healing rituals performed by the tribe's shaman or to a more holistic and humanized treatment vision.

As we said before, this openness to the other is important at each transplant stage.

The same is true, of course, for the post-transplant stage. The conscious use of immunosuppressants requires a particular understanding of the functioning of the human body that makes sense from Western and modern medical paradigms. An average citizen, even if he lives in cities and has some degree of formal education, is often unable to explain the functioning of our immune system. Again: in the following chapter, we will see how the view of disease must be understood in a broader scenario, culturally speaking, so that post-transplant drug treatment must also take this into consideration.

In addition, there is another practical difficulty. The follow-up of the recipient after the surgical procedure can be challenging. Health teams dealing with indigenous populations almost always face challenges regarding infrastructure, resources (human, logistical, and financial), and turnover, so a follow-up routine in isolated communities becomes a Herculean task.

The texts analyzed below explore these issues further and offer solutions for dealing with them.

2.3.2 *Some Challenges to Be Overcome*

Based on the literature, it is time to systematize some of the challenges mentioned above. What are the big knots when considering policies and practices regarding transplants involving native people?

First, the need for effective communication [15, 23–26]. The data from these authors point to an apparently prominent element, but that, in the midst of the need to present high productivity in a short period of time, is often left aside. In fact, because this is the most cited element in the literature, it will be precisely the one to which we will dedicate more space and time in the next chapter.

However, to put it briefly, ineffective communication generates a cycle. By only informing and not educating, uncertainty and disconnection of information are generated, and, consequently, ambivalence about biomedical explanations. Simply put, there is no a priori reason for the native population to believe what they are told by health teams. Their historical experience teaches them that the major ills that plague their lives come, directly or indirectly, from their contact with Western society. And their immediate experience teaches them that their explanations for their illnesses do not make sense from their cosmological systems.

There are several underlying issues there. One of them—again, to be duly explored in the next chapter—concerns the very status of the relations maintained with Western society. The refusal to accept explanations alien to the native cosmogony is often a symbolic refusal to get the very status of the colonized. More than that, communicating and informing is not the same thing as "educating" in a broad sense.

The problem with health actions developed in native contexts (we will see in the next item that there are exceptions) is that they are based on the sender/receiver paradigm. Thus, there would be two poles in the communication: one would provide the information, which in turn would be immediately assimilated by the receiver pole without filters, dissonances, problems, or context. It is as if human beings were a sponge, ready to absorb all information passed to them, candidly and passively.

I like, however, the view of these actions functioning as a "symbolic market," given that the contents tend to be rearranged according to native logics and within their own conceptions of corporality, health, and disease, for example—again, the next chapter will talk about this subject. But in any case, there are several cases that demonstrate how good intentions often do not mean effective health education policies.

Two emblematic stories in this sense, heard by the authors of this book, both from experiences of fighting AIDS in native villages in South America. On one occasion, a health team tried to demonstrate the correct use of condoms in some villages. Everyone here has seen initiatives like this one: a banana is used to demonstrate how to put on a condom, keeping the tip, and wrapping the banana. Weeks later, when revisiting the village, the team found several bananas in the village wrapped in condoms, this being the destination of all condoms left by the team. In their view, the natives imagined that this was a method to preserve the bananas since they saw no direct connection between a banana and a penis. Another story is along the same lines. The team made it clear that condoms can only be used once. When revisiting the villages, they found condoms on clotheslines, drying along with the clothes. When asked why, the natives proudly replied that they had washed them with soap and water. After all, according to the health team, this was the correct way to maintain hygiene in the villages—washing things with clean water and soap—so the connection was immediate. Once cleaned, the condoms would be disinfected and, therefore, clean and ready for use.

In these two anecdotes, whose fault is the lack of effective communication? The easy way out would be to exoticize the attitudes of the natives. Yes, an ethnocentric look forces us to understand native interpretations as "primitive," "backward," or illiterate, for example. Hygiene habits, notions of causality, bodily care: all of this is directly related to the mythological universe of the native peoples, and it is up to the health teams to access this universe in order to understand from what places these people explain diseases, death, and the maintenance of life. Classical Anthropology teaches us, especially from authors such as Marcel Mauss and Claude Lévi-Strauss, that forms of world classification are a fundamental part of any human culture: but this classification makes sense from the point of view of the natives. We have said it before, and it is important to repeat: the fact of being a native community is not a mere detail; their culture is the protagonist in the processes of communication and education, and their participation in each step of the process is very important.

Also, it is important to remember: simply translating materials into the native language does not mean translating them into the native language. It will not be a poster or a workshop in their language that will educate them about what a transplant is, its repercussions, consequences, and processes. These are different chains of causality and temporality, thought from other assumptions and places of interlocution. Imposing biomedical truths without transposing them to native processes is a mistake that is made.

Among those mentioned above, Anderson's text [15] emphasizes how ambivalence in medical explanations is not only a problem in itself but also causes uncertainty and distrust regarding the treatments proposed by medical teams. In other words, not only does ineffective communication not help, it hinders the work. Moreover, this ambivalence toward the explanations provided by health teams runs into other elements to be explored here: the teams' poor cultural competence, problems in the doctor–patient relationship, and different forms of racism.

A cross-cultural environment [15, 28] may highlight another issue quite present in the literature: the low cultural competence of the teams to deal with transplants in native contexts [24, 29, 30]. That said, briefly, a common complaint in the native speeches brought by the literature concerns the little incorporation of indigenous knowledge to the information disseminated about transplants, despite the importance of cultural practices in decision-making regarding transplants [31].

Yeates et al. [30], for example, indicate how "A number of potential mediators may contribute to low indigenous transplant rates, including language barriers, patient preferences, health practitioners' attitudes, and the lack of culturally appropriate patient education programs" although the effective impact of these mediators is scarce in the literature. The fact is that we do not always know where to begin such work in an intercultural context.

Bronislaw Malinowski, for example, seen in contemporary anthropology as one of the founders of the ethnographic method [32], brings a fascinating image at the beginning of his Argonauts of the Western Pacific, published in 1922: "Imagine yourself suddenly set down surrounded by all your gear, alone on a tropical beach close to a native village, while the launch or dinghy which has brought you sails away out of sight ... Imagine further that you are a beginner, without previous experience, with nothing to guide you and no one to help you. For the white man is temporarily absent or else unable or unwilling to waste any of his time on you. This exactly describes my first initiation into fieldwork on the south coast of New Guinea."

This may be the best synthesis of what the experience with Otherness, to which we referred in the Introduction, means. The feeling of isolation, followed by a frightening "Okay, I'm here. Where do I start now?" This book is not intended as a course in ethnography for non-ethnographers, but it is important to make one thing clear: you don't have to be on an isolated island in New Guinea to experience otherness or to pursue cultural relativism. The fact that culture is familiar does not necessarily mean that it is known: that is, anthropological estrangement—the recognition of Otherness—also involves the estrangement of one's own culture in a rotation of perspectives.

This looking from above and from afar, "estranging" oneself means that even if the reader of this book does not work in isolated native communities, several of these inquiries can be displaced to peripheral, poor, or minority communities.

In any case, Malinowski exposes three types of data that one has access to when one decides to understand a population from an ethnographic viewpoint. The first is to understand its structure. This means that one can access relevant information from kinship diagrams or maps, for example. The second is its set of laws and norms: myths and stories. Third, and most important, are the "imponderables of real life." Those small everyday elements that can only be accessed through direct contact: the jokes, the family relationships, the conversation around the campfire. It is about becoming part of the native landscape. Did someone call you to go fishing? Go! Did someone tell you a ghost story? Listen! And so on ... Basically, the task is to get out of the comfort zone and understand the natives. As we mentioned in the introduction, it is about "taking them seriously."

That is, the first step to becoming literate in another culture is to allow yourself to be educated by it. To isolate yourself, metaphorically, from your own cultural, religious, and social assumptions and prejudices and listen to what these people have to say to you. Not out of supposed exoticism or mere professional obligation but because you can actually learn something in the process.

In this way, not only the problem of low cultural competence of the teams is solved, but also the problems in the doctor–patient relationship. Active listening and seeking to learn about important issues in the population you work with make a huge difference from the point of view of everyday relationships. It is, as mentioned before, a set of relationships that can only be understood within the sphere of inter-subjectivity. Even if the doctor sees it as "just another day at the office," the person he/she sees, coming from a completely different sociocultural universe, understands that moment in another way.

These are different views of health and disease [15, 28], but not only that: understanding the native view of health, disease, and body broadens the perspective of healthcare teams. As we have already indicated in the Introduction, and following Stephens' point [29], "Aboriginal people have a different worldview and different beliefs about health that are, at times, at odds with the Western medical model. Aboriginal people have a holistic approach to health that includes wellness in all aspects of their daily life, not just in the physiology of the body. This alternative view of the world influences the way that Aboriginal people approach ill health and critical illness. Although Western medicine may offer a physiological and pathophysiological reason for ill health, Aboriginal people may see beyond these reasons to more spiritual reasons for ill health."

Going further, Culture directly influences not only the perception of the disease but also the therapy and consent process, often deeply impacting the life trajectory of recipients, donors, and their families [10].

An example of this is the various records on Taboos involving organ transplants in native contexts [9, 25, 26, 33–36]. It is important to note, however, that taboos are not always as evident, as the Australian authors mentioned here indicate, about the problems of Maori receiving organs from deceased donors—indeed, almost all the texts that address the issue draw attention to the native preference to receive organs from living donors.

Bongiovanni et al. [9], for example, bring us a synthesis of what would be the spiritual factors that influence the decision to donate or receive organs: "My spirit would feel a special connection with the recipient of my organs"; "I believe that removing my organs from my body damages my body so my spirit cannot pass"; "If I donated my organs, I would worry that my spirit would not be at rest"; "It is wrong for me to donate my organs because I am making myself incomplete"; "I believe that donating my organs will disturb my passing process"; "My religious beliefs do not support organ donation." Drawing a parallel, I recall that Catholics have problems with cremating their own bodies. In their belief, if they are not buried in holy ground, they will not be resurrected at the last judgment. So the solution would not be to scatter the ashes in the wind or in the sea, as is usually done, but instead to bury the urn in a cemetery after the cremation.

The fact is that, as people, we have a spiritual connection to our bodies. Corporality, it has already been said here, is a reflection of an intricate set of social, psychological, and cultural relationships, and the above concerns make this even more evident. Returning to the issue of transplant-oriented health education in native contexts, these aspects necessarily need to be taken into account if minimal compliance and a reasonable post-transplant success rate are to be achieved.

This also involves the readjustment of the language used by the medical teams [15], including informed consent [23]. Here, Devitt et al. indicate something that could go unnoticed on a superficial reading: "It also covered interest in receiving a transcript of the interview and sought instruction on dealing with the audio record at the end of the project. There is a legal requirement for research data to be archived for a set period, but our question specifically addressed cultural issues associated with holding and preserving the audio records of a person who may have subsequently died. In some areas, Indigenous people prefer not to preserve the 'voice' of a deceased person. A total of 125 patients, including 76 Indigenous people, requested their audio interview be destroyed." That is, not only must extra care be taken in relation to the tests and substances taken from native patients, but also care from a culturally sensitive point of view permeates each step, including audio and image capture.

While the consent process differs from one country to another, certain guidelines deserve to be followed whenever possible. In Brazilian legislation, for example, the consent process is seen as a relationship of trust and openness to dialogue and questioning. Consent must be given through adequate channels that respect the individual, social, economic, and cultural characteristics of the person or group to whom the consent is intended, in a clear, objective, and straightforward manner, starting from an interactive and complete communication. In addition, it is up to the patient to withdraw consent at any time, as well as to ask questions and seek clarification whenever they feel the need, seeking an autonomous, conscious, free, clarified, and informed decision. In the case of native communities, consent may need to go through the decision-making sphere of the family of the patients involved, but also, eventually, of the shaman and the community leadership. We have seen here the example of the Maori, and we will work in the following chapter on how some aspects of the native health and body imply consequences for their extended family—that is, a different notion from the Western one of the nuclear family. In other words, it is not just a matter of signing a formal document but an important step in which, once again, culture enters as a relevant variable in decision-making. Again: just translating the document into the native language or making it "simpler" in the case of communities that speak the language of the surrounding society is not enough to account for diversity as a problem.

We drew attention earlier to how different forms of racism also impact, negatively the adherence of native people to organ transplants.

Indeed, several sources call attention to Native perceptions of systemic racism practiced not only by healthcare teams but by the healthcare system itself as a whole [25, 36–38]. Tait (2022), for example, indicates how racism creates stereotypes of Indigenous patients being reluctant, mistrustful, or non-compliant, damaging the relationship with healthcare teams [37]. Incidentally, in the same study, there is an indication that this influence also the exclusion of Native people on the transplant list. In fact, Anderson (2012) also points out how members of the medical team associate the native "poor suitability" with culture and social factors [39]. Opening a parenthesis here, this brings us back to a process of victim-blaming. Let us explain.

We have already indicated here that a common practice is to believe that people, once informed, will adopt practices seen as safer and healthier according to the Western view. However, what we observe is that the population almost always falls back on "unsafe" practices, leaving the question: why, even informed about the risks and dangers, do people continue acting this way? Thus, in almost all health education actions observed in native contexts, health education is taken as a process of modification in the individual's behavior. But more than just changing behavior through information and printed materials, full health education should be perceived beyond the acquisition of knowledge and understanding of specific processes that can lead to health, seen here as the physical, mental, social, and cultural well-being not only of the individual but of the group to which they belong.

This type of intervention is a process of understanding the population for whom the actions are intended, of raising the awareness and politicization of professionals, and of education as a dialogue and exchange of knowledge. Once the correct path has been "taught," any mistake is now the fault of the community to which the supposed educational action was directed. This corroborates visions of infantilization or apathy on the part of those communities as if they act out of disinterest or are incapable of understanding the content that is passed on to them.

More than blaming the victim, in the end, what we see is that the responsibility for his own health is weighed on his shoulders, and the native culture is seen by managers as an obstacle to the application of policies that should be made based on it.

Closing the parenthesis brings us, again, to the issue of racism as an impediment to full transplant policies in native contexts.

Walker (2022) indicates, from a dialogue with Maori leaders, different levels at which racism operates—including explicitly [32]. Racism, the Maori, Walker and his team show us, operates not only in internalized and personal ways but also in institutionalized ways: "Colonization further acts to maintain structural racism and privileges whiteness to normalize racist ideas in media, culture, social systems, and institutions." When discussing health education, we are not restricted to native contexts. We will see below that a diversity-oriented training curriculum in collaboration with native communities is also an urgent need if we want health teams to be minimally prepared to deal with diversity in their daily practices.

Several sources also draw attention to the barriers to inclusion on the list and criteria for organ allocation as problems to be faced [23, 24, 30, 40]. Not only do patients not understand their status on the transplant lists and their inclusion and exclusion criteria, but they also feel little interest from the teams in explaining to them how these processes work—again, the points already indicated in this item regarding ineffective communication, disconnection of information, and racism, in its various levels. However, something else emerges in the literature: how native communities would be harmed by the algorithms of the system that forms the donation lists [28, 30, 37]: minorities would be harmed by factors such as geographical barriers and morbidity profile, among other elements.

The most obvious elements about the difficulties of implementing a broad and systematic transplantation policy in native contexts would reside there: the geographical isolation [23, 25, 27, 28, 34], logistic obstacles [36], low level of education of these populations [30], and low income and lack of medical insurance [27, 34, 37] are only mentioned as important, but without the prominence in terms of weight and complexity of the analyses about the other elements, brought here.

From the literature, the most significant problems concerning transplants in native contexts are related to sociological rather than structural factors. This is also the direction of the solutions presented by the authors analyzed, as we will see in the next item.

2.3.3 *What to Do?*

So far, we have seen how specific problems affect, in different ways, the development of transplant policies in native contexts. We will see in this section that, unfortunately, the proposed solutions do not get the same space in the literature, even if there are, here or there, records of innovative experiences.

We mention here the existence of initiatives such as the *Kidney Check Program*, in collaboration with the *Network Environments for Indigenous Health Research National Coordinating Centre, the First Nations and Métis Organ Donation and Transplantation Network*, and the *Can-SOLVE CKD Network* in Canada. We also mentioned *Hope and Healing* [19], a campaign for organ donation through the use of social networks in the United States, and the *IMPAKT* Program (*Improving Access to Kidney Transplants*) in Australia [23, 28]. Adding to these initiatives is the formation of an initiative described by Tait (2022) of a Think Tank in Saskatchewan, Canada, targeting the northern and more distant regions of that country over the past 3 years [37]. Initiatives like these have several interesting characteristics, which allow for addressing the problems and challenges presented in the previous section.

Let's summarize what these authors write: These are initiatives done in collaboration with native peoples horizontally and collaboratively, respecting their sovereignty. In addition, they seek to involve wise men and native leaders, including confronting preconceived ideas and biases of the health teams themselves—that is, it is a two-way education process. In the same way, by exposing elements of structural racism in health policies aimed at native populations, they also confront the problems that exclude minorities from transplant lists. In this sense, they involve the community based on their notions of health, disease, prevention, and communication: they seek to respect the flow of communication in a native way, almost always based on orality (the importance of this will be discussed in the next chapter).

More than that: such initiatives also bring to the surface native healing and care processes, not only imposing Western biomedical knowledge. In this way, the community is engaged, improving health promotion and impacting the relationships between the healthcare team and the service users.

Finally, these are ideas that seek the formation of networks. Indigenous people, health professionals, leaders, indigenous wise men, managers, and governments are all impelled to dialogue and participate in confronting the barriers presented above. Not only do they learn together, but they share responsibilities. Again, the native communities also have greater governance in managing their health, including financially.

Several authors indicate the importance of developing specific education programs for natives and staff [27, 32, 33, 41]. Such programs would necessarily include not only respect for the native language but also for their ritual processes of spiritual healing. It is important to understand the implications of affirming, as we have done here, that the concept of health in native populations is holistic and that their bodies result from a flow of social relations. In these communities, when a person gets sick, the whole community gets sick together, even in different ways. The care of a necessarily goes through these networks of native psychological and cosmological care, always within their perspectives of health, disease, and cure. The community must be involved in each step of these processes of mutual collaboration and education, developing its vision of transplantation [19, 29, 32, 37]. As Stephens (2007) indicates, "Consultation, communication, and education are the way forward": working together with the native community, integrating their knowledge, and engaging native educators, sages, leaders, and health teams [29].

In addition, one possibility is the use of telemedicine [27, 42] and Artificial Intelligence in more remote or isolated communities.

2.3.4 What Awaits Us

Three elements draw the most attention at the end of this chapter.

First, there are many more reflections on the problems faced in the implementation of transplants in native contexts than proposed solutions or reports of positive experiences. This is not necessarily a surprise since even the texts that consider alternatives for the consolidation and expansion of these transplants must first analyze the challenges faced. However, the discussion reveals common elements encountered in different national contexts—Australia, Canada, the United States, and New Zealand—whose causes transcend economic factors (such as investments in public health) or management (such as hiring human resources, for example). This brings us to a second element.

Surprisingly, the problems pointed out revolve around sociological rather than structural causes. That is: even though the geographical isolation and the low level of education of the natives are obstacles pointed out by some authors, much more space was given to elements such as inefficient communication in multicultural contexts or structural racism, for example. It is evident that structural issues and sociological aspects interpenetrate and feed off each other. The lack of specific training policies for teams that work with native populations, for example, necessarily affects the relationship of the teams with the target audience and the efficiency of communication and health education actions. In the same way, the lack of an anti-racist curriculum, built in collaboration with native leaders and wise men, can be better understood if we consider the little indigenous representation in some decision-making spheres and budgetary decisions. In this way, simplistic explanations such as "the little native access to transplants is justified by the isolation of these communities" need to be put in check, even if we want to address the cause of problems pointed out in this chapter (such as prejudice, the systemic exclusion of minorities from the lists, and the feeling of disconnection regarding the information given by health teams to natives).

Finally, a third element is the lack of systematic reflections on Latin, Asian, and African contexts. We have reserved ample space at the beginning of the chapter to discuss the Latin American absence in this discussion since it is the reality closest to the authors. It would be obvious to affirm that the debate and knowledge about the theme would be expanded by incorporating diverse national contexts in this discussion. This would include, as has been pointed out here, the formulation of networks of researchers whose reflections could work to pressure managers for specific public policies for transplants in native contexts that respect their cultures and social dynamics. However, there is one more relevant point: discussing health care in native contexts leads us to think about institutional racism, welcoming differences, and social inclusion.

As Anthropology teaches us, failing to understand other cultures not only makes us blind to the Other but also short-sighted in relation to ourselves. Thus, by understanding other concepts of corporality, health, disease, and cure, we better understand our notions of person, life, and care. It is in this sense that the next chapter of this book is inserted.

References

1. S.V. Sakpal, S. Donahue, C. Ness, et al., Kidney transplantation in the United States Native Americans: Breaking barriers. S. D. Med. **74**(1), 21–27 (2021) https://pesquisa.bvsalud.org/portal/resource/en/mdl-33691053
2. J. Aramoana, J. Koea, An integrative review of the barriers to Indigenous peoples participation in biobanking and genomic research. JCO Glob. Oncol. **6**, 83–91 (2020). https://doi.org/10.1200/JGO.18.00156
3. O. Giraud, Comparação dos casos mais contrastantes: método pioneiro central na era da globalização. Sociologias **22**, 54–74 (2009)
4. H. Carrie, T.K. Mackey, S.N. Laird, Integrating traditional indigenous medicine and western biomedicine into health systems: A review of Nicaraguan health policies and miskitu health services. Int. J. Equity Health **14**, 129 (2015). https://doi.org/10.1186/s12939-015-0260-1
5. C.E.A. Coimbra Junior, *The Xavánte in Transition: Health, Ecology, and Bioanthropology in Central Brazil* (The University of Michigan Press, Ann Arbor, 2002)
6. A. Diaz-Cayeros, Coalitions, Indigenous peoples, and populism in the Americas. J. Polit. Inst. Polit. Econ. **3**(1), 107–123 (2022). https://doi.org/10.1561/113.00000054
7. L.F.F. Sandes, D.A. Freitas, M.F.N.S. de Souza, K.B.S. de Leite, Primary health care for South-American indigenous peoples: An integrative review of the literature. Rev. Panam. Salud. Publica **42**, e163 (2018). https://doi.org/10.26633/RPSP.2018.163
8. J. Anderson et al., 'Native Americans' memorable conversations about living kidney donation and transplant. Qual. Health Res. **30**(5), 679–692. Available at: https://doi.org/10.1177/1049732319882672. (2020)
9. T. Bongiovanni et al. Cultural influences on willingness to donate organs among urban native Americans. Clin. Transplant. **34**(3), e13804. Available at: https://doi.org/10.1111/ctr.13804. (2020)

10. A.S. Daar, P. Marshall, Aspects culturels et psychologiques de la transplantation d' organes/A. S. Daar & P. Marshall. Available at: https://apps.who.int/iris/handle/10665/56035. (1998)
11. J. Feehally, Ethnicity and renal replacement therapy. Blood Purif. **29**(2), 125–129 (2010)
12. C. Helman, *Culture, Health and Illness* (CRC Press, Boca Raton, 2007)
13. C.W. Park, W.M. Limm, L.L. Wong, Improving living renal transplant: Lessons from a multiethnic transplant program. Hawaii Med. J. **68**(5), 104–108 (2009)
14. J.A. Trostle, *Epidemiology and Culture* (Cambridge University Press, Cambridge, 2005)
15. K. Anderson et al., "All they said was my kidneys were dead": Indigenous Australian patients' understanding of their chronic kidney disease. Med. J. Aust. **189**(9), 499–503 (2008)
16. L.G. DE Souza, R.V. Santos, C.E.A. Coimbra, Age structure, natality and mortality of the Xavante indigenous people of Mato Grosso State, Brazilian Amazon. Cien Saude Colet **15**(Suppl 1), 1465–1473 (2010, July)
17. R.H. Hahn, V.T. Grando, *Sickness and Healing: An Anthropological Perspective.* (Yale University Press, New Haven, 1999)
18. J.C. Rodrigues, *Os corpos na Antropologia. Em: Críticas e atuantes: Ciências sociais e humanas em saúde na América Latina* (Editora Fiocruz, Rio de Janeiro, 2005)
19. R.K. Britt et al., "Sharing hope and healing": A culturally tailored social media campaign to promote living kidney donation and transplantation among Native Americans. Health promotion practice **22**(6), 786–795 (2021, November)
20. M.T. de Souza, M.D. da Silva, R. de Carvalho, Revisão integrativa: o que é e como fazer. Einstein (São Paulo) **8**(1), 102–106 (2010, March 1)
21. C. Guedes, S. Guimarães, Research ethics and Indigenous peoples: Repercussions of returning Yanomami blood samples. Dev. World Bioeth. **20**(4), 209–215 (2020)
22. C. Lévi-Strauss, *The Sorcerer and His Magic. Understanding and Applying Medical Anthropology* (Routledge, London, 2016), pp. 197–203
23. J. Devitt et al., Study protocol – Improving access to kidney transplants (IMPAKT): A detailed account of a qualitative study investigating barriers to transplant for Australian Indigenous people with end-stage kidney disease. BMC Health Serv. Res. **8**, 31. Available at: https://doi.org/10.1186/1472-6963-8-31. (2008)
24. R.C. Walker et al., Experiences, perspectives and values of Indigenous peoples regarding kidney transplantation: Systematic review and thematic synthesis of qualitative studies. Int. J. Equity Health **18**(1), 204–204. Available at: https://doi.org/10.1186/s12939-019-1115-y (2019).
25. G. Lewis, N. Pickering, Maori spiritual beliefs and attitudes towards organ donation. N Z Bioeth. J. **4**(1), 31–35 (2003)
26. E. Bennett et al., Cultural factors in dialysis and renal transplantation among aborigines and Torres Strait Islanders in North Queensland. Aust. J. Public Health **19**(6), 610–615 (1995)
27. S.V. Sakpal et al., Kidney transplantation in Native Americans: Time to explore, expel disparity. J. Health Care Poor Underserved **31**(3), 1044–1049. Available at: https://doi.org/10.1353/hpu.2020.0078. (2020)
28. A. Cass et al., Barriers to access by Indigenous Australians to kidney transplantation: The IMPAKT study. Nephrology (Carlton) **9**(Suppl 4), S144–S146 (2004)
29. D. Stephens, Exploring pathways to improve indigenous organ donation. Intern. Med. J. **37**(10), 713–716 (2007)
30. K.E. Yeates, A. Cass, T.D. Sequist, S.P. McDonald, M.J. Jardine, L. Trpeski, J.Z. Ayanian, Indigenous people in Australia, Canada, New Zealand and the United States are less likely to receive renal transplantation. Kidney Int. **76**, 659–664 (2009). https://doi.org/10.1038/ki.2009.236
31. S.N. Davison, G.S. Jhangri, Knowledge and attitudes of Canadian First Nations people toward organ donation and transplantation: A quantitative and qualitative analysis. Am. J. Kidney Dis. **64**(5), 781–9. Available at: https://doi.org/10.1053/j.ajkd.2014.06.029. (2014)
32. B. Malinowski, *Argonauts of the Western Pacific: An Account of Native Enterprise and Adventure in the Archipelagoes of Melanesian New Guinea* (Routledge, London, 2002)

33. B.L. Danielson et al., Attitudes and beliefs concerning organ donation among Native Americans in the upper Midwest. J. Transplant. Coord. **8**(3), 153–156. Available at: https://doi.org/10.7182/prtr.1.8.3.f89854842113un43. (1998)
34. M. Jernigan et al., Knowledge, beliefs, and behaviors regarding organ and tissue donation in selected tribal college communities. J Community Health **38**(4), 734–740. Available at: https://doi.org/10.1007/s10900-013-9672-2. (2013)
35. M. Keddis, D. Finnie, W.S. Kim, 'Native American patients' perception and attitude about kidney transplant: A qualitative assessment of patients presenting for kidney transplant evaluation. BMJ Open, **9**(1), e024671. Available at: https://doi.org/10.1136/bmjopen-2018-024671. (2019)
36. N. El-Dassouki et al., Barriers to accessing kidney transplantation among populations marginalized by race and ethnicity in Canada: A scoping review Part 1 – Indigenous communities in Canada. Can J. Kidney Health Dis. **8**, 2054358121996835–2054358121996835. Available at: https://doi.org/10.1177/2054358121996835. (2021)
37. C.L. Tait, Challenges facing Indigenous transplant patients living in Canada: Exploring equity and utility in organ transplantation decision-making. Int. J. Circumpolar Health **81**(1), 2040773. Available at: https://doi.org/10.1080/22423982.2022.2040773. (2022)
38. R.C. Walker et al., "We need a system that's not designed to Fail Maori": Experiences of racism related to kidney transplantation in Aotearoa New Zealand. J. Racial Ethn. Health Disparities (Internet) [Preprint]. Available at: https://doi.org/10.1007/s40615-021-01212-3. (2022)
39. K. Anderson, J. Devitt, et al., If you can't comply with dialysis, how do you expect me to trust you with transplantation? Australian nephrologists' views on indigenous Australians' "non-compliance" and their suitability for kidney transplantation. Int. J. Equity Health, **11**, 21–21. Available at: https://doi.org/10.1186/1475-9276-11-21. (2012)
40. A. Cass et al., Renal transplantation for indigenous Australians: Identifying the barriers to equitable access. Ethn. Health **8**(2), 111–119 (2003)
41. J. Devitt et al., Difficult conversations: Australian Indigenous patients' views on kidney transplantation. BMC Nephrol. **18**(1), 310. Available at: https://doi.org/10.1186/s12882-017-0726-z (2017)
42. K.A. Barraclough et al., Residential location and kidney transplant outcomes in Indigenous compared with Nonindigenous Australians. Transplantation **100**(10), 2168–2176. Available at: https://doi.org/10.1097/TP.0000000000001007. (2016)

Chapter 3
Native Corporeality and Organ Transplantation

3.1 Working with Other Cultures: What Now?

Specific discussions do not work well in the abstract. As we write this book, we know that many people prefer a pragmatic approach to health care. After all, to say that Otherness imposes challenges on health care may sound strange to those who have learned in their academic training that biomedical knowledge has truth value in itself. However, even if we take modern and Western medicine as a paradigm, the care for culturally differentiated populations necessarily involves understanding the Other and their notions of body, disease, health, and death.

This chapter will present some concrete examples of how this happens in practice. Our first idea was to bring several examples from around the world of how Culture and Embodiment are interconnected, but after some more thought, we reconsidered this strategy. Bringing in various people in this chapter would be exoticizing their perspectives. We would turn the book into a patchwork quilt without the necessary space for deeper reflection. The fact is that any perspective of generalization would necessarily be omissive or superficial.

Instead, we will base our discussion on our concrete experience as public health professionals in Central Brazil and the Amazon, where we have over two decades of work. If we have practical experience of how challenging health care in an intercultural context can be, sharing some of our reflections can lead us to interesting conclusions. A deep dive from an empirical situation will thus be our starting point.

E. R. Fernandes, A. K. Nobrega Cavalcanti, *Organ Transplantation and Native Peoples*, SpringerBriefs in Public Health,
https://doi.org/10.1007/978-3-031-40666-9_3

3.2 Talking About Health in Intercultural Contexts

3.2.1 Two or Three Words on Intercultural Health Communication in Native Contexts

The first theme, addressed by the authors mentioned in the previous chapter, is communication and education in health. The Shavánte are an indigenous people of about 18,000 who inhabit the central portion of Brazil, in the southern Brazilian Amazon. They will be, to a large extent, our guides in thinking about some of the well-established practices in the field of health care with native peoples. If we want to approach transplants in native peoples with these communities, what mistakes do not need to be made?

A few years ago, talking with a Shavánte leader, we heard a story about how the amoeba associates with certain foods consumed in the city and then infects children in the villages. Hearing this, it is clear how their ontology understands specific processes of what some diseases are. This complicates the simplistic perspective from which Western categories of health and illness are transmitted, in a linear fashion, to native peoples. Actions of prevention, communication, information, and health education necessarily go through understanding the native cosmology, history, and politics.

As Lopes da Silva [1] points out, anthropological research on educational processes involving indigenous peoples should "make use of procedural analysis, contextualization and historicization of the situations analyzed, awareness of the multifaceted character of identities, social groups in interaction and political strategies in interethnic contexts, overcoming rigid oppositions such as the classic 'traditional' x 'modern,' 'indigenous society' x 'surrounding society,' 'native (pure, authentic) institutions' x 'exogenous institutions', 'pure' x 'acculturated' Natives."

The overcoming of these oppositions and this view of processes (that is, of understanding points of view as a becoming) also permeates the interventions in the health field.

The processes implied in this relationship should be interpreted if taken as "signs of relations with otherness" [2]. This means that the subjectification of these mechanisms (sickness, death, bodily care, etc.) goes through the understanding of the indigenous socio-cosmological universe, the social reproduction of this universe, and the relations that indigenous people maintain with themselves and with the Other, going far beyond a historical contingency, a manifestation of identity, an interethnic relationship or inevitability in the scope of relations with the State.

An example of this is some reflections on Health Education in native contexts. Jean Chiappino, writing about the production of didactic material for Native communities south of the Venezuelan Amazon, proposes the articulation between Western medicine and the shamanic ways of those native peoples (Yanomami, Yekuana, and Kuripako, among others). According to him, encouraging the community to participate in the decisions that affect their health allows them to leave behind an interventionist model that promotes passivity in those populations.

Otherwise, interest is lost, and there is resistance to health promotion from a biomedical perspective. The way out is, then, how to combine the knowledge of different cultures, seeking outlets that are linguistically and culturally bilingual [3].

Seen in this way, one can infer that biomedical information is only one of the variables in a complex field of interethnic interactions. Often, and we will see this later, resisting treatments imposed by Westerners symbolizes a resistance to Westerners themselves and what they represent.

We propose not a "native tradition" in opposition to biomedical knowledge but knowledge (ontologies) *in relation*, one to the other: the axis of analysis, in this sense, would be located in the production of shared meanings.

This has practical effects. In actions of education, communication, or health information in native contexts, sometimes the initiatives are understood as a space of Western knowledge in relation to indigenous knowledge; sometimes as a Western knowledge re-signified by the Natives. What we propose here is an approach that considers the contents passed on by the professionals involved in relation to indigenous ontology.

The relative lack of impact of these actions, however well planned and elaborated, is not due to a lack of intelligence on the part of the target audience or its culture. The problem is precisely how initiatives are placed in this panorama, from the outside in, from the top down. The conceptual shift that we propose here is a view of these processes from the inside in, by understanding that native culture is more than a "simple detail" that can easily be circumvented by the literal translation of the instructions that would characterize those materials as if they were aimed at a context other than that of native peoples.

In the same way, the images used in the leaflets and posters allow us to see that, at all times, the message is directed toward—and from—a stereotypical indigenous person.

Tassinari [4] gives us an example based on an article by Jean Jackson about the Tukano in Colombia, about school teaching and shamanic practices. Jackson narrates the attempt to create workshops that would work as "shaman schools," a project idealized by an anthropologist in the 1980s, seeking to "rescue the indigenous traditions." In the article analyzed by Tassinari, the author (Jackson) examines why such initiatives failed. The first point concerns the very nature of shamanic knowledge and how it is transmitted—which would hardly fit into the Western model of knowledge transmission. The second point Jackson draws attention to is the internal diversity of the Tukano, with shamanic knowledge in the Vaupés region of Colombia being by no means homogeneous.

Finally, the shamans were seen as holders of legitimate knowledge, receiving money from the government to transmit some information that they, in the end, refused to share in that environment.

In what contexts is knowledge passed on, by whom, and to whom? We almost always take questions like these as a detail without realizing their importance throughout planning health actions in native contexts.

3.2.2 "The Virus Is the Health of the Disease"

The Shavánte wise men we spoke with are almost unanimous in saying that they have heard the health teams talk about certain diseases but that, in general, the natives say that these things are not valid.

Some of these conversations occurred during the outbreak of diseases such as the H1N1 outbreak in 2009 or diseases caused by worms, bacteria, viruses, or (according to the health teams) "lack of hygiene." We were interested in knowing to what extent the information about the new flu reached the elders. What is noticeable in the Shavánte accounts and explanations about diarrhea, amoebas, viruses, bacteria, and worms is that they tell us something about how they see us: our concepts of health and disease reach them but are transformed into an ontology of otherness from the Shavánte cosmogony.

According to some interlocutors, some diseases would have arrived with the Westerners: verminous, for example, would exist in the villages due to consuming sugar, milk and coffee, olive oil, and "cold" (industrialized) food. Children, in turn, would get sick from sleeping with their bellies uncovered, with "diarrhea entering them" at night. But interesting to note: it is the White man's diarrhea, which is harmful and can kill, caused by the food that the Shavánte eat when they go to the city.

Even though the Shavánte's explanations of the origin of worms, diarrhea, viruses, and bacteria vary widely, in almost every conversation we had with them, two things did not change: (1) Westerners' diseases are different, and they kill; and (2) Even though these diseases have different origins (some say that the amoeba comes in the white people's food, as in the case above, others—we will see below—maintain that the Shavánte are already born with amoebas, viruses, and bacteria and that they develop, after consuming the white people's food), their causes are almost always related to the consumption of foods such as olive oil, sugar, salt, alcoholic beverages, rice, etc.

The testimonies heard almost always contain good points for thought in this sense:

INTERLOCUTOR 1

What were diseases like before the arrival of the whites?

At the time before contact, there were some diseases caused by witchcraft. Still, there were no diseases that came from outside because people ate only food from the savanna, always passing the medicines in the body, so there were not so many diseases. Only spells made against the enemy.

Is there any difference between the diseases of whites and Natives?

Several kinds of diseases kill us. The disorders are different. The Natives come weak, so there is almost no death. But today, the conditions come stronger, killing. Today the Natives no longer do some rituals to avoid these diseases because there are practically no more elders—who are dying more because of the white people's illnesses.

(...)

Have the doctors explained about these (white) diseases?

They talk a little. Viruses exist, for example. We drink contaminated water, then we catch the virus and get diarrhea and vomiting. (...) In the old days, we didn't know what the virus was, drank contaminated water, and didn't get sick because we took medicine from the bush. This medicine we used to take in the old days doesn't ruin life, but the white man's medicine does. The white man's medicine already contains the virus for us.

And how do Natives get diseases from white people?

Why do we get sick today? The amoeba and the virus already exist in our bodies from an early age. Since childhood, we have been eating coffee, candy, and olive oil, which gives us headaches. So, through this food from outside, a worm grows, making us sick; it provides us with stomach pain, diarrhea, and vomiting. This happens; it is the amoeba.

INTERLOCUTOR 2

What were the diseases like before contact?

In the old days, there was no such thing as a bad cough, no such thing as a headache. But there was such a thing as a disease through a spell. This disease was more potent. And the sick person transmitted it to the other. But stronger diseases, such as vomiting, fever, and headache, did not exist.

Do the media explain diseases? About viruses, bacteria?

The virus is the life of the disease. The virus that is with us, giving us illnesses like vomiting and diarrhea ... the virus exists. In the past, the virus was found in stagnant water; when man hunts and drinks water, there is already the worm. But in the past, we didn't know about the virus, but we know that the virus passes through the water. Then, when you drink it, this little bug comes with the water. A cousin drank water to quench his thirst, and then he had a bug, giving him stomach pain and ruining his stomach, making him vomit all day; then he died because he drank still water, and the virus came. The virus is also in the food that is not covered. When we eat cold food, what we eat there is already contaminated with the virus, and then it transmits the vomiting, the headache, the fever. The mosquito comes and lands on it and leaves its feces.

INTERLOCUTORS 3 AND 4 (interviewed together)

What were the diseases like before contact?

In the old days, diseases such as fever, cough, and flu did not exist. The only thing that existed was the amoeba, but there was a shaman for this. When someone had diarrhea, the shaman would give them medicine, which would be solved. There was a shaman for stomach aches and vomiting. The shaman knew all this. The illnesses that used to exist in the past, the stronger ones, were caused by the shaman to catch each other's relatives. These diseases cannot be cured because these diseases come from the shaman. That is why there is no medicine for them. (...) The diseases of the Natives and the Whites are different. With the Natives, sometimes you get sick, but with the Whites, when you get them, you die. Today there is white medicine for diseases, but this medicine contaminates us and brings death (...)

... and why do white and Native diseases transmit differently?

The father transmits it to his children. For example: if the father eats the meat of a rhea, the father gets diarrhea, and his eyes become similar to the rhea's, leading to death. Because of the things that the father eats, it is not the father who will get sick; it is the children who will die through the transmission of what the father eats. The armadillo, the snake, all of this is forbidden to kill and eat, to preserve our children. (...) So-And-So has already killed the snake, his children were small, and he didn't even know that it would bring diseases to his children. His children almost died. (...) The amoeba exists, but it is transmitted by the parents, even. When the father has amoeba, the child is already born with the father's amoeba. Nowadays, sometimes it is transmitted too because we get used to alcoholic drinks, sweets, and salt, but the amoeba existed before the contact with white people.

In this regard, Trostle [5] writes that the diffusion of epidemiological data undergoes significant symbolic transformations, from which individuals reinterpret scientific data according to their own experiences and systems of meanings. However, I warn the reader that it is not just a matter of stating that they take the facts that are informed to them through health education actions and reinterpret them through their culture. It is a matter of affirming, here, that such facts that are passed on to them through health education policies permeate both health education among the Shavánte; and the Shavánte perspective on corporality—and, in the cases addressed, on conception; and their cosmology, myths, and historical perspective.

As we could see from the interviewees' statements, their ideas based on what is exposed by the health teams are much closer to an indigenized perspective on the contents, related to notions such as corporality, substance, and conception than necessary to viruses and bacteria, as the health professionals postulate in their educational materials, lectures, etc.

It seems to make much more sense to take viruses, bacteria, amoebas, and worms from a Shavánte cosmology than to "import" *ipsis literis*, the cosmology of Westerners. This leads us to another question: even if health education in the Shavánte does not take place in what we could call a "shamanic context of knowledge transmission" (referring here to the discussion raised by Jackson among the Colombian Tukano), what we notice is that the Shavánte cosmogony tries to "shamanize" them to, in this way, account for the Other.

3.2.3 What Makes a Native?

What makes a native? Far from being ambiguous nonsense, the question is more than pertinent and can be unfolded into two others: what makes a Native define himself as such (i.e., as a person) and what shapes him, literally (i.e., what endows him with physical corporeality).

Both questions converge on the other that the way the Native perceives himself socially will affect how he conceives his material existence, just as how he sees his own body—and others see him and he sees others—will have consequences in his construction as a social person.

Such questions briefly pointed out here have been the object of (good) anthropological reflections for some time now. In his classic article, *The Techniques of the Body* [1934], Marcel Mauss was one of the first to call attention to the fact that the body is an eminently social substrate. The same Mauss, in another classic text, *A Category of the Human Spirit: the Notion of Person* [1938], also tried to demonstrate how the vision of the self is not innate but socially constructed. Since then, several authors have worked on issues related to the body, death, illness, and healing, under the most diverse perspectives and in different societies—the excellent book by Cecil G. Helman [6], for example, is illustrative in this sense.

In Social Anthropology, however, especially in studies of ethnology, such concerns gained impulse, especially after 1979, in a small text entitled *The Construction of the Person in Brazilian Indigenous Societies* [7]. We will situate below some interesting consequences of this article, but before that, we will synthesize it into what may be of more interest to us for this work. The primary hypothesis of the text is that "The body, as we Westerners define it, is not the only object (and instrument) of society's impact on individuals: the naming complexes, the ceremonial groups, and identities, the theories about the soul, are associated in the construction of the human being as understood by the different tribal groups. It, the body, affirmed or denied, painted and pierced, guarded or devoured, always tends to occupy a central position in the vision that human societies have of the nature of the human being. To ask, thus, about the place of the body is to start an inquiry into how the person is constructed."

It is about trying to interpret the language of native corporality. When a native says that his body is different from the Western body (as we will see later), or that pigs see themselves as people and we as jaguars, or even that his shaman is seeing the spirit that is causing an illness, what does this mean?

Important texts on the subject date from this period, such as *Os Mortos e os Outros*, by Manuela Carneiro da Cunha (1978)—on the eschatology of the Krahò Indians, a Gê group from northern Brazil—and *Nomes e Amigos: da prática Xavante a uma reflexão sobre os Gê*, by Aracy Lopes da Silva (1980). Such a moment can only be understood in the light of what was occurring in Anthropology at that time. The reanalysis of classic ethnological models is based almost always on situations located in Africa or North America in the context of the Time and Space in Lowland South America seminar in 1976: "Our analytical problem is that of phrasing order when we know the order is there but have no language through which to express it" (Actes of XLII Congrès des Américanistes, Paris, 1977).

The theme of corporeality gained new life in the mid-1990s with the discussion on Amerindian Perspectivism (mentioned in our Introduction). This moment was so impactful that it became part of the so-called "ontological turn" in world anthropology.

In 1996, the articles by Tânia Stolze Lima and Eduardo Viveiros de Castro published in the journal *Mana* (entitled *The Two and Its Multiple: Reflections on Perspectivism in a Tupi Cosmology* and *Cosmological Pronouns and Amerindian Perspectivism*, respectively), studies on Amerindian corporality gained new momentum. Before entering the perspectivist terrain, however, it may be worthwhile to take up some aspects of the Shavánte.

For them, *A'uwẽ* means "people" or "person," while the whites are called *waradzu*, a term traditionally translated as "foreigner" or "stranger." Based on these assumptions, being human would necessarily imply being Shavánte. The problematization of these categories will serve as a starting point for some reflections.

While human (*Homo sapiens*) for Westerners is an innate category, one becomes a person through a process in which one becomes a member of a given society through one's existence as a member of a given culture. As we will see below, the Shavánte classification system is more prospective than the Western one, reinforcing even more the idea of becoming a process of transformation of the being.

In 2000, one of this book's authors interviewed a Shavánte chief about a myth related to the Shavánte's theft of fire, once possessed by the jaguar. It is a story so well known and essential among the native peoples of the Americas that it is the myth with which Levi-Strauss opens his classic "The Raw and the Cooked."

At first glance, the jaguar is more humane than the natives, which at the time, seemed problematic. According to this story, in the old days, the jaguar was the one who possessed the secrets of medicine, body painting, fire lighting, music, etc. The natives, on the other hand, "ate rotten wood," insects and larvae, and did not know how to cook food, paint themselves, etc. After several incidents, one of the natives managed to steal the fire, and the jaguar, to take revenge, has since eaten those it meets.

In the interview, the cacique patiently tried to explain the apparent contradictions in this story.

Here is part of the dialogue:

- Tell me something: is a jaguar a person?
- Yes, jaguars are people. Jabiru, anteater, peccary, tapir ... They are all people ...
- Are A'uwẽ?
- Yes, they are A'uwẽ.
- But, don't you eat these animals? How can you eat them if they are people?
- We eat, but it's because they have already turned into a bug.
- But didn't you say they were A'uwẽ?
- But they are A'uwẽ ...Get it?

In this dialogue, an important notion is noted: perspective, from which one is or is not, according to a relational system articulated by the Indians. Thus, in what "position" would the whites find themselves in the Shavánte classification systems? I suggest here that the white is classified according to a selective and relational logic, in which something specific belongs to a particular category in a specific context.

As I thought, operating with dichotomous categories, this would not fit into the classificatory logic according to which, through a simple syllogism: if A = B, and B = C, then A = C.

However, this does not seem to work with the native perspective categories: maybe part of A, under a particular condition and from a certain point of view, is part of B, under a specific situation and from a certain point of view. A portion of this part of B, under particular conditions and from some perspective, can be equal to an amount of C, under a specific situation and perspective, and so on. More than that, even the concept "equal to" can be read as "derives from" or "varies from."

In any case, it is clear that: (a) contextualization will determine what one is or is not; and (b) such a system presupposes heterogeneity, in which there are intermediate categories between a "white" pole and an "indigenous" pole, in a kind of gradation.

To say that native peoples manipulate the identity categories of the world around them is nothing new. Vilaça [8] has masterfully demonstrated this in relation to the Wari. For this author, "the notion of the body as a site of difference is not limited to interspecific relations. Society is conceived as being constituted by bodily aggregates of various levels, and its boundaries are so variable that it becomes difficult to speak of society."

Thus, when it is said that, for the Shavánte, our bodies are different, it is necessary to understand that, first, it is not a question of a genetic or biological conception of the body; second, it is a question of a fluid and relational position, where one is, and is not, and, third, this relational position will vary according to the degree of personhood one confers on the other.

Let us return, then, to the concept of Amerindian perspectivism. This concept was first elaborated by Eduardo Viveiros de Castro in 1996. It was summarized by the author on that occasion as the "conception, common to many peoples of the continent, according to which the world is inhabited by different species of subjects or persons, human and non-human, who perceive it from different points of view" [9]. This notion was the fruit of the author's reflection on several references in Amazonian Indian cosmologies to "an indigenous theory according to which the way humans see animals and other subjectivities that populate the universe—gods, spirits, the dead, inhabitants of other cosmic levels, meteorological phenomena, plants, sometimes even objects and artifacts—is profoundly different from the way these beings see them and see themselves."

The implications of this concept for this work are countless, among which is a better understanding of identity categories through body language: "The set of ways and processes that constitute bodies is the place of emergence of difference" [10].

What are the effective practices in the time and space of health education? What representations involve its traditions by its actors and the public? What new products and social actors do such an enterprise produce—books, primers, indigenous health agents, etc.? What effects does it have on internal political and social dynamics? What is the relationship, if any, between traditional knowledge and the knowledge transmitted in these new contexts and its forms of transmission?

Some texts point in exciting directions in this sense. One of them, written by Viveiros de Castro, brings an example very similar to those already mentioned here: "A mission teacher [in the village of] Santa Clara [where the Piro people live, in the Peruvian Amazon] was trying to convince a Piro woman to prepare food for her small child with boiled water." The woman replied, "If you drink boiled water, you get diarrhea." The teacher, laughing mockingly at the response, explained that common childhood diarrhea is caused precisely by drinking unboiled water. Unperturbed, the Piro woman replied, "Maybe for the people in Lima, this is true. But for us, native people here, boiled water gives us diarrhea. Our bodies are different from yours." [11].

What analysis does the author propose on the subject? Let us see: "The anecdote of the different bodies invites an effort to determine the possible world expressed in the judgment of the Piro woman. A possible world in which human bodies are different in Lima and Santa Clara—in which the bodies of whites and Indians must be different. The argument that "our bodies are different" does not express an alternative, and of course mistaken, biological theory or an imaginary non-standard objective biology. The Piro argument manifests a non-biological idea of the body, an idea that makes issues such as childhood diarrhea untreatable as objects of a biological theory. The argument states that our respective 'bodies' are different, meaning that the Piro and Western concepts of the body are divergent, not that our 'biologies' are different. The Piro water anecdote does not reflect another view of the same body, but another concept of body, whose underlying dissonance in its 'homonymy' with ours is precisely the problem" [11].

As we have seen here, it can be said that the natives' concept of corporality is more perspectival than the Westerners', giving rise to their constant reinterpretations.

One of the messages that the stories we have seen so far concern the question of what it is to be human: when the amoeba reaches the Shavánte through non-indigenous food when the white man's diarrhea kills, and the Shavánte does not ... Such stories tell us something about what it is to be a native in a position, relational and perspective, in relation to the white. This brings us to the question with which this text was opened: what makes a Native?

Being Native, as the self-denomination of these peoples allows us to perceive, is equivalent to being a Person. In other words, this means one has the bodily properties, dispositions, and aptitudes necessary to develop relationships with fellow beings; the myths and old men's stories teach us that physical properties change because the human condition changes. The bodies become different because their position in society changes.

The white body is not only other, but its difference in relation to the Shavánte body marks the border of sociability. From here, two consequences above all call attention: (a) there is a difference between Natives and whites that is relational and prospective and can be continuously managed; and (b) if, on the one hand, such a border serves to establish differences between the distinct forms of sociability, on the other hand, it serves to mark relations of substance and kinship that shape the native person.

In our Shavánte example, the first point seems to be the most evident: we may not be Shavántes ourselves, but since it is a relational and perspectival system—as we have already mentioned—it is possible that, in certain situations and from specific points of view, we become Shavántes—that is, we become indigenized.

The second point (relations of substance and kinship) is a little more complicated, and I mention it here by complementing the argument. For the Gê-Speaking tribes (among which the Shavánte), father, mother, brothers, and sisters "basically share the same bodily substance, when one of them eats those hot foods, or foods of the color of fire (red), or the meat of fish with sharp teeth, the effects (the heat, the pungency) will be felt by the sick relative" [12]. As other authors point out, "according to this logic, parents, children, and kin are bound by relations or bonds of bodily identity that go back to the conception that ultimately reduces the body to a single essence—semen, creator, father" [13].

Among the Shavánte, there are plenty of examples of food taboos: during pregnancy, parents are not allowed to see, touch, hunt, or eat armadillo meat, as the child's body may undergo some changes (including eating dirt when crawling), similarly, parents are not allowed to eat tapir meat, lest the child's body shortens; beans will make the child's skin darker; etc.

However, the Shavánte body is not only made of the father's semen. There is the soul, which arrives at the body through the action of the creator spirits of *Dañimité*.

In addition to the soul, there is the name. Although a Shavánte may have up to six names in the course of his life (each name corresponding to an essential phase of his life), the child is born without a name and remains so for some time. "As soon as he is born, a child (of either sex) cannot receive a name. He has to grow up a little. The name is too heavy a load for his fragile, "soft" body that will eventually become ill until he dies. Therefore, the child must remain nameless while it is small. When he grows up, and his body becomes "harder" and more resistant, the name will not cause him illness" [14].

Now, the name is the public and social marker of the person: once again, corporality appears as a marker of sociability: the child is not yet a person, and therefore, he does not have a name and, because of this, his body does not "hold" a name. In addition, Coelho de Souza points out that besides the substance and the name, the skin is a constituent element of the Gê tribes' body. Particularly in the Shavánte case, this is noticeable differently than in northern Brazilian Gê-Speaking tribes, such as Kayapó and Xikrin [15].

If the body paintings are highly elaborate among the Kayapós, among the Shavánte, the paintings are graphically simpler. However, elements such as bracelets and necklaces assume a fundamental role: the painting establishes the ritual function of the person, the one who makes his ornaments situates him socially, etc. A hitherto unexplored aspect in Shavánte ethnographies concerns that the skin serves as a transition zone between the exterior (the name, kinship, public and ritual sphere) and the interior (the substance, semen, blood). If we think from this angle, it is not surprising that corporality functions as the language of Shavánte relations with otherness: in the stories we have seen, the changes operate not only in physical corporality but in the human form of being, denoting, with this, the change of status in the relational personhood conferred to the Other.

The conclusions and implications of the ideas presented here are numerous. Among these, I first highlight the importance of understanding Native corporality as an idiom of the indigenous relationship with the world around them. As we have seen, how they perceive themselves says a lot about how they perceive the world. Moreover, we were able to draw attention to a logic of construction of the person that is different from ours: a selective, relational, and perspective logic, which gives fluidity to body forms and world construction. Going a little further, from the stories presented here, we realize how fluid bodily properties are: this relational body is where differences emerge.

3.2.4 Untying the Knots

A few things can be said from the elements we brought in this chapter. The white body is not only different, but its difference from the Native body marks the border of sociability. The bodies become different because of their position before society changes.

In the example given throughout these pages, the fundamental question is the following: if our bodies and diseases are so different, why do our conditions contaminate them? The answer we propose next is that this is precisely the reason.

The body works here as an alterity marker: we are the "other," the potential enemy. In a certain way, some processes can be better interpreted if we observe how the natives appropriate these new experiences and spaces and that such appropriation is reflected in their thinking and their relations with otherness.

This process of symbolic production that we call "culture" is highly dynamic and covers processes that, to our eyes, are contradictory.

Here the parallel to what Zourabichvili writes about the Deleuzian idea of *devenir* (becoming) becomes evident: "The relation thus mobilizes four terms and not two, divided into intertwined heterogeneous series: x involving y becomes x′, while y taken in this relation with x becomes y" [16].

It seems plausible to state that, depending on the context, intermediate categories are circulating between the white and native poles. We need only remember that since cultures are in permanent transformation, our vision of the "other" and, consequently, of ourselves is also changing.

Finally, since native peoples show flexibility in dealing with the speed with which the new conditions posed by the contact situation arise, we must understand how native peoples position themselves in the face of this historical process. However, their response to the new situations will be sought within the field of possibilities in their society.

Thus, when it is said that, for the Natives, our bodies are different, it is necessary to understand that (a) it is not a question of a genetic or biological conception of the body; (b) it is a question of a fluid and relational position, where one is, and is not; and (c) this relational position will vary according to the degree of personhood one confers on the other.

Bruce Albert writes about the Yanomami that "The retrospective logic and structural operations of these processes of mythological extension and reconfiguration are very characteristic of the analogical creativity that allows Amazonian shamans to constantly update their group's mythology in the light of the novelties and contingencies of immediate history" [17].

One of the motivators of this research was a quick conversation we had with a fellow researcher by chance while we were walking to the bus stop. When she heard we were trying to understand health education in native areas, she answered laconically that the research would be unproductive. According to her, we would conclude that "the Natives are losing their traditional vision to assume ours." What we realized was just the opposite. As Carneiro da Cunha writes in a short article on Shamanism, the process of shamanic translation "is not only a task of tidying up, of keeping the new in old drawers; it is a matter of rearranging, more than of tidying up" [18].

Thus, we argue that structurally speaking, there is "room" in Native ontology to inscribe what is presented to them. However, and we hope to have demonstrated this, one can only understand how such processes work when we take the relationship between history, cosmology, and indigenous corporality.

This brings us to another point.

To view the disease in a monolithic way would diminish the possibility of seeing essential aspects of the disease: "Physical condition and economic hardship influence susceptibility. Work, diet, water source, poverty, activity patterns, and residence influence exposure to infectious agents [...] Sick people belonging to different social groups have correspondingly different levels of access to health services, and these health services provide varying levels of quality of care" [5].

It is important to emphasize here that it is not just a matter of enumerating the social determinants of a given disease or of reducing the entire process to a mere culturalist reading. Methodologically and conceptually linked to Epidemiology, Anthropology provides a broader and complementary view of health and disease processes.

From the moment that anthropological and epidemiological practices are seen as complementary, it is possible to understand how culture interferes with self-care patterns, in the perception of the environment and space, in the systemic process of perceiving the health and disease process, and even to evaluate the suitability of health intervention policies for specific populations (especially the so-called minorities and disadvantaged groups) from the "inside" view of these cultures of these processes.

Insights into how such groups perceive different health practices, public policies, "scientific" knowledge, etc., are essential not only from an analytical point of view but also politically since health policy managers need to recognize such plurality.

In this sense, the study of native people is quite adequate. If, for example, there are norms that foresee the respect for the medical practices of culturally differentiated groups, and in a certain way, the structure of services takes place as if this fact were verified in practice, what occurs is an imposition of Western clinical and biomedical approaches, given as the only ones possessing a status of "truth." As Garnelo and Langdon point out, "The training of health professionals, based on biologism and individualism, does not enable them to distinguish an individual idiosyncrasy from a group pattern of behavior forged from specific cultural productions. Usually, the interaction between professionals and patients excludes the sociocultural references that allowed them to perceive the latter as agents of transformation of the social environment" [19].

There remain, here, some questions—one consequence of the other: in the first place, to what extent is this procedure of the health teams a reflection of a State ideology in terms of the standardization of its macro policies? Second, concerning the public policies applied to native peoples and such process of uniformization of public policies, referred to above, are there possible interconnections between this practice of cultural homogenization and the scientistic ideals that guide the formation of the idea of the nation since the beginning of the twentieth century? I remind the reader that race, nation, and ideology were the trinomial behind normalized eugenicist and racist practices in the first half of the last century. Everything indicates that the answer to both questions is "yes."

The direct consequence of these processes is the lack of dialogue with indigenous knowledge, and a relative imposition, as previously stated, of biomedical practices as holders of the monopoly of truth. From the moment such rules are imposed on the native peoples, the symbolic inequality between the indigenous peoples and the State increases, and, through this inequality, this imposition is legitimized in a vicious circle.

However, during this process, new knowledge and wisdom are created. Here again, comes the contribution that anthropology can give to epidemiological studies of minorities: the understanding of health policies and services, health and disease processes, and self-care. To see native peoples as mere subjects in this context is not only naïve but also deprives them of the power of agency in the face of the policies directed at them.

3.2.5 An Important Note Before We Move On

By now, the reader should have realized something important, but in case he has not, it is vital to note this before we move on to our Conclusion.

You may work with people from the African savannas, with aborigines in Oceania, or even with traditional peoples from European regions. So, a lengthy explanation of the cosmology of specific people from the Brazilian savannas may not touch you. It is essential to make clear that the Shavánte were used as an ethnographic example because they are a tribe with whom we have done decades of research. They were our guides in a more general and ambitious journey. This allows us to bring more profound knowledge about their socio-cosmological universe and develop some points of our argument.

This means that the observations made so far are not restricted to the Shavánte or the Gê-speaking peoples of Central Brazil or even to Brazilian native peoples. They do not even concern only indigenous peoples or even traditional or original populations. Peripheral peoples generally do not conform to Western biomedical explanations of the processes of illness and death. We use here several ethnographic cases—the Piro, Tukano, Yanomami, Kayapó, e.g., in which this is evident. Bodies function as social markers of difference, and how different cultures deal with their corporealities teaches us a lot about their knowledge concerning Westerners (and, consequently, to their biomedical explanations). Bodies are meeting points of social flows, and an interdisciplinary reading of these dynamics is necessary if we want to intervene epidemiologically in these populations.

However, as we said above, this is not restricted to communities in which radical alterity is given. The difference between a white Westerner and any member of an indigenous people is evident: their ways of speaking, thinking, and acting operate from distinct cultural logics. They must be understood from different interpretative keys. In other words, when considering transplants involving people from these Cultures (our problem in this book), issues such as consent, corporality, family, affection, and social representations of the disease start from entirely different paradigms. The medical team knows this and almost always acts, even if within its limits, keeping this in mind.

Even if the health team does not have a minimal understanding of what Cultural Relativism is, care is taken to make a rotation of perspectives, from which one seeks to understand the culture of the Other in terms of the Other. This is the basic idea of Relativism. Thus, relativist practice is applied, even without knowing the anthropological concept. An interpreter is sought, and eventual health education actions—including with the medical team—take into account some evident cultural differences; eventually, the help of a leader, elder, or shaman is sought to help conduct these processes.

But what happens when this radical Otherness is not put in place? Supposedly accessible information is not always manageable, even for people with the same cultural background. Working with a population that is not necessarily native or traditional, for example, speaking the same language, having the same skin color, and living in urban areas, does not mean that the processes behind organ transplants are necessarily transparent or understandable.

Even if you have never seen a native person in your life, chances are you have had experiences in the field or in your practice where there was an unbridgeable gap in communication. This is not a normative grammatical barrier but a difference between social representations of how corporeality operates. This was the main reason we brought up the health education issue in this chapter, starting with an extreme example.

The pandemic revealed a massive ignorance of how modern medicine explains viruses and immune systems. Covid-19 has made this clear in many countries. The misinformation made all kinds of epidemiological flat-earth appear: there were several fake news about the disease, prevention, and treatments. In the interior of Brazil, for example, some made videos claiming that looking at a welding spark would "kill the virus." All sorts of fictitious treatments and unproven medicines have been used worldwide, e.g., drug cocktails, miracle cures, homemade teas, and vitamins.

All of us in the scientific community were appalled at the level of misinformation and the enormous number of lives that could have been saved if communication in science and health had been done more quickly and effectively. These were often not people living in caves, deserts, and remote forests: many were middle-class citizens of Western countries with access to an unprecedented flow of information, which could have been filtered through basic science information. Where did we go wrong?

Communication processes, information, and health education are the first step to considering transplants seriously. However, we forget this or think it is just a matter of linguistic translation. We take Western science as something given and consolidated, as if all people speak "science," which is invalid. In other words, on the one hand, we have to reinvent ourselves. Here we try to make clear that there is no opposition between "traditional" knowledge and "modern" knowledge, but rather that they overlap and reconstruct each other. It is much more a relational logic and perspective than necessarily something watertight. This means, in practice, that professionals who speak "science" must be more flexible and learn to speak other languages, thinking from keys that take Alterity seriously, breaking with an ethnocentric vision of Health.

We have seen how the health of blood relatives often affects the health of newborn children and how some cultures understand the Western world as responsible for pathologies more than bacteria and viruses. Contagion, prevention, the systems that make up our bodies ... all of this must be relearned when dealing with different cultural and social realities. This should also have been a lesson of the Covid-19 pandemic, going beyond the issue of native peoples.

How do you tell a person without access to sanitation in his home to wash his hands constantly or to buy alcohol to sanitize them? How do you ask a poor informal worker to maintain social isolation at home when he has to work to bring food to his family? How do you tell a mother who lives in a small house that eventually contaminated people must stay in isolation when many times there aren't even doors between their rooms if their home has rooms? There is a need to access this universe to act consciously and effectively in health care without blaming the victim for the disease or exoticizing the victim's social or cultural universe.

In the field of transplants, the discussions in this chapter teach us several powerful lessons. The main one is that the difference between native and Western bodies tells us about the evils of colonialism and how native peoples deal with Otherness.

Bodies are not only flowing of symbolic and subjective relations; they are syntheses of the native philosophy of the connection to the Other. If one wants to take transplants seriously in intercultural contexts, not only physical bodies must be understood in other terms, but the indigenous relationship with its relation to Westerners must be understood, too, and several times, the resistance to transplant initiatives in people from native peoples, and also opposition to Western intervention in native life itself—such as the story of the Piro woman, told here. In this way, we can only turn on the key of intercultural sensitivity and exercise our active listening. Non-Western therapeutic itineraries go through a holistic view of one's body and a complex view of its history and relationship with its Others. This includes not only the whites but the dead, the animals, the inbred relatives, and the kindred.

Often a collaborative approach can mean successful treatment. This is not to say that medical transplant teams should become social scientists, learn their native language, and become experts in the cultures of the people they work with. However, they (we) can do no wrong in trying to understand how the Other deals with these issues to move away from a strictly interventionist paradigm.

References

1. A.A. Lopes da Silva, Educação indígena entre diálogos interculturais e multidisciplinares: introdução, in A. Lopes da Silva, M.K.L. Ferreira (org) *Antropologia, História e Educação – A questão indígena e a escola* (FAPESP/Global Editora/MARI, São Paulo, 2001)
2. E. Viveiros de Castro, O. Calavia, M. Coelho de Souza, C. Fausto, B. Franchetto, C. Gordon, C. Lasmar, T. Stolze Lima, M.T. Pinto, A. Vilaça, *Transformações Indígenas: os regimes de subjetivação ameríndia à prova da história. Projeto Pronex CNPq-Faperj* (Núcleo Transformações Indígenas, Rio de Janeiro, Florianópolis, 2003)

3. J. Chiapino, *Participación comunitária al control de la salud: experiência de producción de documentos didácticos en Amazonas venezolano*, Anales Nueva Época, n. 5 (Instituto de Estudos Ibero americanos, Göteborg University, Göteborg, 2002)
4. A. Tassinari, Escola indígena: Novos horizontes teóricos, novas fronteiras da educação, in A. Lopes da Silva, M.K.L. Ferreira (org) *Antropologia, História e Educação – A questão indígena e a escola* (FAPESP/Global Editora/MARI, São Paulo, 2001)
5. J.A. Trostle, *Epidemiology and Culture* (Cambridge Press, Cambridge, 2005)
6. C. Helman, *Culture, health and illness* (CRC Press, Boca Raton, 2007)
7. A. Seeger, R. Da Matta, E.B.V. de Castro, *A construção da pessoa nas sociedades indígenas brasileiras*. Boletim do Museu Nacional, Série Antropologia, n. **32**, 2–19 (1979)
8. A. Vilaça, O que significa tornar-se Outro? Xamanismo e contato interétnico na Amazônia. Revista Brasileira de Ciências Sociais **15**(44), 56–72 (2000)
9. E. Viveiros de Castro, Os pronomes cosmológicos e o perspectivismo ameríndio. Mana **2**(2), 115–144 (1996)
10. E. Viveiros de Castro, *A inconstância da alma selvagem* (Cosac & Naify, São Paulo, 2002)
11. E. Viveiros de Castro, O Nativo Relativo. Mana **8**(1), 113–149 (2002)
12. A. Seeger, *Os índios e nós: estudos sobre sociedades tribais brasileiras* (Campus, Rio de Janeiro, 1980)
13. M.S. Leite, R.V. Santos, C.E.A. Coimbra Junior, *Alimentação e nutrição dos povos indígenas no Brasil* (Editora Fiocruz, Atheneu, 2007)
14. A. Lopes da Silva, *Nomes e Amigos: da prática Xavante a uma reflexão sobre os Jê*, Tese de Doutoramento em Antropologia Social (São Paulo, FFLCH/USP, 1980)
15. M. Coelho de Souza, *O traço e o círculo: o conceito de parentesco entre os Jê e seus antropólogos*, Tese em Antropologia defendida no Museu Nacional (Universidade Federal do Rio de Janeiro, UFRJ, Rio de Janeiro, 2002)
16. F. Zourabichvili, V. Goldstein, *O vocabulário de Deleuze* (Relume Dumara, Rio de Janeiro, 2004)
17. B. Albert, A. Ramos (Orgs.), *Pacificando os brancos: cosmologias de contato no Norte-Amazônico* (UNESP-Imprensa Oficial do Estado, São Paulo, 2000)
18. M. Carneiro da Cunha, Pontos de Vista Sobre A Floresta Amazônica: Xamanismo e Tradução. Mana **4**(1), 7–22 (1998)
19. L. Garnelo, E.J. Langdon, *A antropologia e a reformulação das práticas sanitárias na atenção básica à saúde* (Editora Fiocruz, Rio de Janeiro, 2005)

Chapter 4
Conclusion: Merging Horizons

A first reaction when we come across the literature on this subject is one of dismay.

Several of the texts analyzed here make clear the contradictions of thinking about transplants in native contexts. There are few financial and human resources available, relatively little interest from academia, and issues concerning medical training with little room to think about cultural diversity. In addition, native peoples face a longer waiting time on transplant lists, even though they have alarming rates of diseases that imply impairment of their organs. Native peoples in the United States have three times higher mortality from kidney disease and high rates of diseases such as diabetes compared to minority groups [1–7].

On the other hand, as Sakpal et al. indicate, the "Federal funding for the Indian Health Service (IHS) is low: $3,851 per NA/AI in 2017, which compares with $10,348 per capita health care spending in the U.S, and $8,602 per person in U.S. federal prisons in 2016" [8]. There you have, apparently, a scenario of an apparent wasteland.

Apparently, in fact.

Stephens (2007) draws attention to the fact that indigenous people who need to undergo hemodialysis almost always have to move away from their families, having to travel to large urban centers far from their communities. This makes them often prefer to die of kidney disease at home. A transplant would not only mean that these lives would be saved, but it would also mean the quality of life for the patient and their families [9].

Despite being considered a minority group, indigenous populations constitute a significant portion of the population in many countries and are almost always placed at the margins of national health systems. As we have exposed here, many of the challenges faced in the health care of indigenous populations are also barriers to be overcome in peripheral or poor people. Caring for these vulnerable social groups necessarily means being open to the Other.

E. R. Fernandes, A. K. Nobrega Cavalcanti, *Organ Transplantation and Native Peoples*, SpringerBriefs in Public Health,
https://doi.org/10.1007/978-3-031-40666-9_4

Thus, we have seen how each step of the transplant, from the consent process on, needs to consider the patient's cultural specificities. The way the exams are done, the destination of images and body fluids: all this needs to be discussed and dialogued. We also saw how forming knowledge networks that allow the health teams to access more effective communication with the indigenous wise men and their leaders is also necessary. This directly affects the professional team's relationship with the patient, his family, and his community.

We discussed at length in Chap. 3 how diseases are perceived from other perspectives of corporality, health, and personhood. Even how the community sees its contact with the surrounding society becomes fundamental to the effectiveness of health education actions, for example. It is important that the team listens to and involves the social group they work with as part of the patient's therapeutic itinerary, always keeping in mind that the vision of health is more holistic than the Western one.

The community gets sick when one of its members gets sick, just as the whole community participates in its healing processes.

In addition, processes that lead to the blaming of culture or that normalize racism in dealing with the patient must be avoided at all costs, which also means a process of educating health professionals themselves. The process of taking the culture of the Other seriously implies, precisely, being guided by their anguishes, doubts, and concerns. As we have seen here, these are other visions of temporality and causality, including cosmological elements at work. This means understanding that an idea of linear causality, based on Cartesian logic, can make sense from a modern, Western scientific paradigm. However, not all social groups share this worldview.

As we said here, even in Westernized populations, medical paradigms are sometimes not fully understood—the recent Covid-19 pandemic, as we mentioned here, was proof of this. Communication is only effective when it is thought outside the logic of the sender and receiver, and the context and channel become as important as the message itself.

Culture is a symbolic way of communicating something. The disconnect in health education in native contexts occurs by imposing symbols whose meanings are not shared. A poster on the wall of a hospital or a message sent to a patient can have countless meanings depending on who reads it, in what context, who wrote it, etc. An example of this has already been mentioned: when we were campaigning to increase breastfeeding in a native Brazilian group, the indigenous mothers were outraged. According to them, white people did not breastfeed but gave them powdered milk. We, the health teams, wanted to impose breastfeeding because "that's how the animals do it." It was not about health but an arena of symbolic conflict in which powdered milk became a paradigm of humanity.

Similarly, the example of mother Piro, mentioned here, of workshops using condoms … There would be many examples of good intentions carried out by a properly trained technical staff seeking solutions to serious health problems in native contexts. But, still, without any effectiveness.

Working with health in native contexts implies merging horizons of perspectives. As we said in the Introduction, this does not mean abandoning medical paradigms or eschewing protocols. However, incorporating some of the native social and cultural dynamics into treatment can be the difference between life and death for patients. Dombá Suyá, for example, could only perform the transplant because she had the help of her shaman. His transplant team had the sensitivity, in the anthropological sense of the term, to understand that their healing processes necessarily passed through native knowledge. As we have pointed out here from the beginning, Culture matters.

The realization that different cultures do not mean unequal cultures is the first step to effective action in native and peripheral contexts, in general.

References

1. J. Anderson et al., 'Native Americans' memorable conversations about living kidney donation and transplant. Qual. Health Res. **30**(5), 679–692. Available at: https://doi.org/10.1177/1049732319882672 (2020)
2. T. Bongiovanni et al., Cultural influences on willingness to donate organs among urban native Americans. Clin. Transplant. **34**(3), e13804. Available at: https://doi.org/10.1111/ctr.13804 (2020)
3. R.K. Britt et al., "Sharing hope and healing": A culturally tailored social media campaign to promote living kidney donation and transplantation among Native Americans. Health Prom. Pract. **22**(6), 786–795. Available at: https://doi.org/10.1177/1524839920974580 (2021)
4. M. Jernigan et al., Knowledge, beliefs, and behaviors regarding organ and tissue donation in selected tribal college communities. J. Commun. Health **38**(4), 734–740. Available at: https://doi.org/10.1007/s10900-013-9672-2 (2013)
5. H.R.L. Wiley et al., Kidney transplant outcomes in Indigenous people of the northern Great Plains of the United States. Transplant. Proc. **53**(6), 1872–1879. Available at: https://doi.org/10.1016/j.transproceed.2021.05.003 (2021)
6. E. Bennett et al., Cultural factors in dialysis and renal transplantation among aborigines and Torres Strait Islanders in North Queensland. Aust. J. Public Health **19**(6), 610–615 (1995)
7. K.E. Yeates, A. Cass, T.D. Sequist, S.P. McDonald, M.J. Jardine, L. Trpeski, J.Z. Ayanian, Indigenous people in Australia, Canada, New Zealand and the United States are less likely to receive renal transplantation. Kidney Int. **76**, 659–664 (2009). https://doi.org/10.1038/ki.2009.236
8. S.V. Sakpal, S. Donahue, C. Ness, R.N. Santella, Kidney transplantation in Native Americans: Time to explore, expel disparity. J. Health Care Poor Underserved **31**(3), 1044–1049 (2020). https://doi.org/10.1353/hpu.2020.0078
9. D. Stephens, Exploring pathways to improve indigenous organ donation. Intern. Med. J. **37**, 713–716 (2007). https://doi.org/10.1111/j.1445-5994.2007.01500.x

Index

A
Alterity, 4, 5, 10, 14, 16, 17, 19, 48, 50, 51
Anthropology, 3, 9, 10, 27, 28, 34, 42, 43, 49, 50

C
Comparative studies, 14, 16, 22, 23
Corporality, 1, 6, 9, 21, 26, 29, 34, 42, 43, 46–48, 50, 56
Cosmologies, 1, 6, 38, 42, 43, 45, 48, 50

D
Diversity, 1, 2, 10, 13, 16, 23, 30, 31, 39, 55

E
Education in health, 38

H
Health of indigenous peoples, 38, 50
Health policy, 23, 32, 49, 50
Health services, 14, 15, 49, 55

T
Transplantation, 1, 2, 9, 10, 13, 14, 16–18, 21–23, 31–33

E. R. Fernandes, A. K. Nobrega Cavalcanti, *Organ Transplantation and Native Peoples*, SpringerBriefs in Public Health,
https://doi.org/10.1007/978-3-031-40666-9